ENVIRONMENTAL PLANNING

FOR UTILITY CORRIDORS

by

ROSS T. NEWKIRK
Associate Professor
School of Urban & Regional Planning
Faculty of Environmental Studies
University of Waterloo
Waterloo, Ontario

ANN ARBOR SCIENCE
PUBLISHERS INC
P.O. BOX 1425 • ANN ARBOR, MICH. 48106

230 Collingwood, P. O. Box 1425, Ann Arbor, Michigan 48106

Library of Congress Catalog Card No. 77-92593
ISBN 0-250-40223-8

Manufactured in the United States of America

PREFACE

This computer-based planning system was developed to help optimize environmental resource allocations when locating continuous utilities. It is oriented to identifying corridors for potential utility development over study areas from 2,000 to 100,000 square miles in extent. When established environmental assessment and routing methodologies were reviewed, few were found to directly apply to the total utility corridor routing problem. The new system is a unified and flexible planning tool which integrates all major corridor identification phases. Features include automated data acquisition, development of a study data bank, impact factor assessments, computer-drawn maps, impact factor combination, and a multistage routing system. These features combine to provide the identification and assessment of a set of alternative corridors. Practical application of part of the system to routing a high-voltage transmission line is surveyed.

Major advances achieved by this system include not only the system structure itself but also a sampling method for direct data input from maps to the study data base, a Cascade Algorithm for combining ordinal impact assessments (accommodating threshold conditions), a corridor routing phase based on graph contraction and a modified Dijkstra algorithm. While environmental utility routing is the main emphasis, some of the methodology developed may find future application in the development of regional data banks, impact factor models, voting behavior, digital picture enhancement, computer backboard wiring plans and the layout of printed circuits.

Ross T. Newkirk
Waterloo, Ontario

Dedicated to Rosemary
and to the memory of
A. I. Johnson
Dean of Engineering Science
University of Western Ontario

Ross T. Newkirk is currently Graduate Officer and Associate Professor, School of Urban and Regional Planning, Faculty of Environmental Studies, University of Waterloo. He holds a BA in economics, MSc in computer science and PhD in engineering science from the University of Western Ontario and has been an honorary lecturer in geography and computer science there as well as assistant director of the Social Science Computing Laboratory.

Dr. Newkirk's areas of academic specialization include resource analysis and planning (environmental impact assessment), computational methods (modeling simulation, computer science, graphics) as well as spatial and regional analysis.

The author of many technical publications, Dr. Newkirk is President of the Canadian Regional Science Association, and is a member of the Regional Science Association, Association for Computing Machinery, IEEE and Society for Computer Simulation.

CONTENTS

CHAPTER 1

PERSPECTIVE ON THE ENVIRONMENTAL ROUTING PROBLEM

INTRODUCTION

Planning and design for the placement of major utilities has always been a complex economic and physical problem. This is true in particular when a utility must follow a continuous route. Railways, highways, electricity transmission lines and pipelines share the requirement of following continuous routes and simultaneously involve substantial capital investment. The engineer's planning model must attempt to route the utility to maximize profit by minimizing construction cost and eventual operating costs. This has given rise to the development of near linear routes between source and destination with deviations from linearity being due, in large part, to avoidance of physical impediments which would increase construction costs. Such location of utilities is often coupled with destruction of scenic areas, conflict with residential or commercial land uses, and major alteration of drainage systems and land use patterns. Such alterations in ecological, social, economic and physical characteristics of an area are called project "impacts."

Where some of these impacts have been sufficiently severe, remedial action by governmental agencies is required. This action can range from physical alterations in the neighborhood of the utility to development of alternate services or amenities to replace those destroyed by the utility either during construction or operation. In many cases, these actions are required a long time after initial construction due to slow development of side effects. Society may pay a significant real cost directly attributable to the development of the utility. It is unlikely that these costs will have been included fully in the engineer's original estimates. The addition of these "social costs" to the construction and operations costs normally considered could reduce project profitability sufficiently to render a project unattractive. Opponents of projects are quick to seize upon such costing discrepancies and use them to call for government intervention. In the face of increasing pressure from project opponents and financial outlays for remedial actions, governments are now requiring proponents to consider what is called "environmental impact" in their project planning.

In the United States the act to establish A National Policy for the Environment (U.S. Congress 1970), commonly called the National Environmental Protection Act (or NEPA), requires impact statements to be filed for all "environmentally significant" projects. While the act itself involves various curious features (Kreith 1973) requiring proponents themselves to decide if impact is significant enough to require an impact statement, it has given rise to increased attempts at assessing environmental impact. In Ontario, the Green Paper (Ontario Ministry of the Environment 1973, Plewes and Young 1973) of the Ministry of the Environment outlines various possibilities for legislation requiring environmental statements; recently specific legislation has been promulgated. Utility companies and governmental agencies proposing new utilities in Ontario are required to submit impact statements before Ministry approval is granted. Lucas (1973) outlines Canadian legislation which requires projects to file some kind of "environmental impact" statement. In many cases, only general requirements are listed and methodology details are unspecified. This is not inconsistent with the general situation in the United States.

An examination of the summary in Dickert and Domeny (1974) of U.S. federal and state guidelines shows that while some attempt has been made at outlining what actions require the filing of an impact statement, little has been specified detailing the data, weightings and analysis methods to be used. In the absence of these details, proponents are developing their own methodologies. As a result dissimilar methods may be applied in a class of similar projects making objective comparisons difficult. In Canada, the approach of Ontario Hydro is typical of the response of utilities to this situation. Of six major environmental studies it commissioned during 1972-73 for the location of major 500-kilovolt (kV) transmission lines, no two studies shared the same methodology or similar data base.

Often, environmental impact review is performed fairly late in the planning process–long after the basic framework of the project has been set (Dickert and Sorensen 1974, Fisher and Davies 1973, McNeil 1973, Twiss 1974). Some researchers have proposed that environmental impact assessment be imposed as a "side condition" after the initial feasibility study is complete (Nijkamp and Paelinck 1973). Three possible outcomes of project review may be identified (Ontario Ministry of the Environment 1973; Plewes and Young 1973): approval to proceed, approval to proceed subject to certain conditions, or refusal to permit further proceeding. Accordingly, a proponent could face cancellation or costly major revision of a project after major financial outlays for design and planning have been made. Even if a proposal is granted permission to proceed or to proceed subject to specified conditions, a substantial delay in the project may be experienced due to the review process. This is particularly true if insufficient or unclear analyses are presented. The increased costs resulting from lengthy delays include lost business, and increased costs of construction. An increase in one or both of these "costs" could serve to render a project uneconomical. It is strongly argued by McNeil (1973) that sufficiently detailed and accurate environmental assessment must be included at all levels in the planning process. Fisher (1973) and Lucas (1973) agree; in addition they observe that early review facilitates proper public involvement and improves the credibility of the resulting assessment. It is evident that a planning method is required which will provide sufficiently detailed analysis capabilities for the utility routing problem to accommodate environmental impact assessment.

NATURE OF THE UTILITY ROUTING PROBLEM

Utility routing is a most difficult spatial resource allocation problem. Multiple sources and destinations may have to be joined by continuous routes over long distances. "Control points" may need to be accommodated. These are points or regions, which for system integrity reasons, must be passed through or avoided. Seemingly small changed in local topography, resources or land use may cause serious engineering problems as well as environmental impacts. Malisz (1972) observes that impact discontinuities are the dominant feature in the routing problem.

The problem is further complicated due to the necessity to locate optimal routes over very extensive tracts of land or study areas. A "study area" is a tract of land which contains both the source and destination of a route and all the land over which a route may be permitted to pass between source and destination. A routing problem for a short pipeline could involve a study area 50 by 20 miles; for a long pipeline similar to the proposed Mackenzie Valley pipeline, a study area could be 2,000 miles long by an average of 60 miles wide. Study areas may run from 1,000 to 200,000 square miles in extent. A major difficulty is to analyze accurately the discontinuous impacts over these large tracts of land.

It is necessary to evaluate possible alternate optimal routes since a number may exist. In addition, it may be appropriate to change the analysis to accommodate alternate technologies. This, in turn, could give rise to additional alternate routes. Given that various alternate routings may be possible some method is required to perform an objective comparison between them. This requires extensive tabulations and mapped display of project impacts.

A difficulty which is often encountered while selecting a route is to trade off the cost of a longer route against avoiding regions of higher project impact. It may not be possible to locate a route of any length which avoids all regions of major impacts. In the event that a certain amount of impact is to be tolerated at isolated points along a route, it is necessary to develop a methodology which will help determine which impact should be accepted to minimize total impact of the route as well as detail the significant costs of alternate routes.

A general overview of the routing problem has been given. The problem may be re-stated in Operations Research terms as follows: it is an extension of nonlinear programming. In the traditional nonlinear optimization problem one attempts to locate a point in multidimensional solution space which finds the global maximum (or minimum) of an objective function (or surface) subject to many constraints. Various methods are outlined in Cooper and Steinberg (1970). The routing problem is slightly more complicated by the necessity to locate a path (*i.e.,* a series of points) from a source to a destination in the multidimension solution space such that the average path elevation is minimized (or maximized) subject to constraints and path length. A further restriction is that the path must be continuous and (possibly) its altitude must not exceed some arbitrary constraint value. Since environmental data change (often discretely) from one subregion to another, it is evident that there may be a number of optimization problems—one for each subregion.

EXISTING METHODOLOGY: COMPLICATING FACTORS

Various methods have been developed to perform environmental impact assessment; many are discussed in the following chapter. Few have been developed to solve the environmental routing problem. Accordingly, work in environmental impact assessment of routing projects is at a very basic level. It is necessary to develop some new methods borrowing from existing methods where possible. While each existing method has specific problems of greater or lesser degree, all face a number of general problems. These include: appropriate involvement of public and governmental concerns, knowledge and attitudes; sufficient precision in dealing with discontinuous phenomena; the requirement to perform similarly structured analyses over wide ranges in project scale; flexibility to accommodate emerging government policy and the results of project reviews by adjusting study components and parameters quickly; delineation of alternative development strategies; assessment of the No Development alternative; and the basic problem of the judgments applied by the assessors in assigning values, weights and ranks to items in a study.

In the environmental impact area there is not a universally accepted methodology. The situation is further complicated by the paucity of standards for many types of possible projects. For example, there are very few environmental standards published concerning routing high-voltage electricity transmission lines.

FEATURES OF A NEW METHODOLOGY

The author has developed a planning system which integrates most significant steps involved in performing a detailed environmental impact analysis and associated least impact route finding. In conjunction with a group of colleagues, the author has applied most system features to the task of locating a 500-kV electricity transmission line. That application is discussed in Chapter 11. The author's system itself is a process of applying, sometimes iteratively, a series of steps. Steps include determination of the major environmental factors for consideration through data collection, data bank building, developing factor assessment modules, combination of assessments, and actual route development. Included in the process is the generation of mapped and tabulated partial and summary results as well as provision for input of public and governmental concerns, knowledge and attitudes. Data acquisition to support data bank development is automated by using digitizing equipment and special computer programs. Factor assessment modules are developed based upon the data bank and implemented by special computer routines. The combination of factor assessments, assessment mapping and route selection are performed by a series of computer procedures.

Major technical features of the planning system can be outlined:

1. An "open ended" data bank is developed which can contain any well ordered combination of nominal, ordinal or interval scaled data. Data scale can be arbitrarily selected to meet the requirements of a project and may support different levels of recording precision appropriate for the diverse variables included. New data items may be added to the bank with minimal difficulty during the development of a study.

2. Automated data acquisition and effective data certification are used to develop the data bank. These techniques are implemented using low-cost digitizing equipment and nonspecialist operators while maintaining a high degree of data accuracy.

3. Full generality of impact factor models is permitted (subject to data availability constraints). The number and complexity of these models are arbitrarily selected for specific applications. They can be refined and expanded with relative ease as required during the duration of a study.

4. Various methods may be tried for the combination of impact factor assessments. A new algorithm is provided to process threshold conditions.

5. A multistage procedure is available for the development of alternative least-impact routings. Included in this procedure is the provision of well defined arbitrary weighting and assessment schemes. The procedure includes tabulations of alternative route impacts.

6. A generalized computer mapping system which produced good quality publishable maps is included. This facilitates both internal and external review of partial and final results of studies.

Functional implications of planning system operations provide some desirable features:

1. Ease of data input from arbitrarily scaled map series encourages the development of a data bank of "original" level data. This can reduce the professional judgmental bias in the data bank which would result from input of interpretations.

2. Explicit environmental assessment modules must be developed utilizing the data bank to produce impact predictions. This tends to force the development of rules which are applied consistently and uniformly across the study region.

3. The flexibility available in developing assessment module structure as well as the parameters associated with route development permit the iterative adjustment of the study to best accommodate the divergent views of the proponent, the public and government.

4. Public and governmental input can be added to a study in a variety of ways. Areas of special concern to the public can be added in mapped form to the data base. Values developed from public surveys can be used in impact analysis modules. System output of maps and tabulations can be used to permit effective public review during the development of a study. System flexibility to accommodate refinement of module parameters adjustments to the data base encourages the direct incorporation of concerns, criticism and new information obtained during the external (*i.e.*, public and governmental) review process.

PLANNING SYSTEM MAJOR COMPONENTS

The author has designed both a planning system and a set of computer procedures to implement it. During the development of this planning system a number of important components were developed. They are summarized here and described in more detail in later chapters.

A Totally Integrated System

The system integrates all study steps from impact factor and data base development through to actual routing analysis. This is done in a modular way to facilitate iterative development of study segments. This helps achieve a system which has the flexibility needed to react to emerging policy and standards. The author knows of no similar system which permits the total integration of all planning phases and permits full analysis flexibility. Chapter 4 describes the system in general; Chapters 5 through 10 describe specific components. The integrated and flexible nature of the planning system is the basis of its practical value.

Automated Data Base Development

A computer procedure to interpolate grid area data directly from source map sheets has been designed. At minimal cost, mapped input can be added to a computer data bank at appropriate recording precision. The author knows of no planning system with such a facility. This is an important feature in making the system practical to use.

An Advance in Combining Assessments

A new Cascade Algorithm has been developed to permit combination of multiple assessments into one total assessment on a stacked or cascaded threshold basis. It avoids problems associated with attempting to use weighted average combinations of ordinal data values. This algorithm provides good resolution and helps to simplify the route-finding process. The algorithm is discussed in detail in Chapter 9. A particularly important feature is its ability to generate threshold *No Development* conditions from a number of moderate impact assessments. There are possibilities for additional applications of this algorithm in telemetry signal processing, problems in remote sensing data analysis, and voting coalition problems.

Routing Analysis

A multistage route identification and analysis procedure has been designed and implemented. Using graph contraction, a modified Dijkstra route-finding algorithm, route straightening, and corridor assessment it effectively processes large problems. The procedure guarantees the location of least-impact routes and provides alternates. In addition, it removes path irregularities consistent with path impact restrictions. Flexibility is included to declare in a continuous scaling low, moderate and high penalty areas in addition to "No Go" or "No Development" areas. There are possibilities for additional applications of this procedure in developing optimal wiring plans for communications and computing equipment as well as optimal layout of printed circuit boards. The reduction of the routing problem using graph contraction appears to be a useful contribution to route finding in large networks. This feature of the planning system is discussed fully in Chapter 10.

CHAPTER 2

EXAMINATION OF EXISTING METHODOLOGY

Standard methodologies may be conveniently divided into general environmental impact assessment methods and path-finding methods. There are many different techniques used or proposed to perform the former. The latter is relatively poorly developed with respect to its integration with environmental impact assessment. Most path-finding development has been associated with network analysis, critical path studies and operations research. Very few attempts have been made at combining general environmental impact assessment methods and path finding. Notable exceptions are the GCARS system by Turner (1970, 1972), its application by Hausmanis (1971) and the work by Krauskopf and Bunde (1972). It will be observed that many of the impact assessment methods are not easily restructured to include route finding.

ENVIRONMENTAL IMPACT ASSESSMENT

The purpose of an environmental impact assessment as observed by Dickert (1974) is to *identify* a set of impacts which can be anticipated, and to *specify* the range which these impacts may take including their spatial dimensions and time frame. In addition, an assessment should *evaluate* and *communicate* the relative trade-offs possible between various alternatives. Nine major methods may be identified for general environmental impact. They include "Expert Committee" assessment, checklists, matrix methods, benefit/cost analysis, descriptive land unit analysis, input-output analysis, overlay techniques (both manual cartographic and computer-based), information system transformation analysis, and analysis by mathematical surface approximations.

Expert Committee Assessment

"Expert Committee" assessment is by far the most unspecific method. It simply consists of assembling a number of academic or technical experts from various fields and requiring them to write a report describing the various impacts they identify. Areas or routes of minimum impact are developed by consensus. The results of such a study are strongly influenced by the subject area backgrounds of the study team, relative dominance of the investigators and the realtive incisiveness of the studies conducted by individual researchers. The quality of such assessments is not consistent from one study to another. A good example of this approach is the Nanaimo Port Study (Environment Canada 1974).

Assessment by Checklists

Checklists have been used for some time in environmental assessment. They grew out of a desire to establish some common ground to be covered by "expert committee" assessments. The lists are defined by some governmental agency or interest group identifying, for specific projects, lists of environmental conditions or factors which should be considered and reported upon. Such lists may be very general (Resources for the Future 1972) or very specific for a certain class of projects. An example may be found in Dickert (1974). Environmental assessments based on checklists usually contain a narrative which reviews the impact or lack of impact associated with each item on the list. While such lists may be helpful in enumerating what impacts might be reviewed, they provide no basic methods to identify the impact *per se* or to rank such impacts when detected. A danger implicit with checklists is the propensity of assessments not to consider those possible impacts which are not on the list but could have project importance.

Assessment by Matrices

Matrix methods are developed from checklists. These methods can be viewed as multidimensional checklists (Dickert 1974) where functional interrelationships may be identified. Often, only the most fundamental, readily obvious, or first-order relationship is indicated. In some cases, a dimension may be added ranking the importance of such interrelationships. Matrix development is usually performed only on a regional summary basis for a project and its alternatives.

The most widely known matrix is described in the Geological Survey Circular 645 by Leopold *et al.* (1971). In this matrix, the indices in one dimension list a set of possible functions which might be detected in a project, and indices of the other dimension list a set of possible impacts which could be associated with functions. Assessors inspect the matrix and indicate an impact ranking (from 1 to 10) for each association. An extra dimension has been added for indicating the relative importance of each such association. No methods are specified by which these rankings may be determined or by which total impact is to be evaluated from the resulting matrix.

Ross (1974) describes an interesting extension of the matrix approach using matrix powering techniques (Berztiss 1971) borrowed from graph theory and network analysis. He develops an Environmental Component Interaction Matrix and an Interaction Disruption Matrix to study biological interdependencies beyond the first order. These matrices are used to assess the direct (or first-order) impact and secondary impacts of various alternatives proposed for a project. An application of this method may be found in the Nanaimo Port Study by Environment Canada (1974).

The use of matrix assessment is very widely supported. Bishop (1974) points out that assessment matrices may help identify important components to be considered in benefit/cost studies. Dickert and Sorensen (1974) propose a tiered matrix development and claim that matrix methods help identify important interrelated effects. Hill and Tzamir (1972) use matrix methods and scaleogram analysis to evaluate alternatives. They claim improved results over benefit/cost analysis and transformation functions. However, these applications tend to be single summaries over large regions and are not helpful in locating routes within a study area.

Benefit/Cost Analysis

Benefit/cost analysis has become increasingly important in environmental impact analysis. It was originally developed in the early 1960s in the United States Office of the Budget where various governmental projects were reviewed and not approved for federal funds expenditure unless the project had a net benefit. Anticipated benefits were identified, "valued," and compared to expected discounted costs of project development. It is immediately clear that the selection of so-called benefits from a project and the selection of items identified as costs as well as their respective weightings or valuations totally determines the final result of any benefit/cost study. The appropriate benefit and cost components to include are often a matter of debate only to be overwhelmed by disagreement concerning the values to be assigned to each. Andersen (1974) provides a good review of benefit/cost analysis applied to environmental impact studies. He observes that a major problem is the failure to evaluate correctly secondary and side effects of a project. Accordingly, project costs are understated giving rise to support of questionable projects, the proponents being very careful to identify all conceivable benefits.

Benefit/cost analysis has usually been based on gross national income (GNI) effects as benefits. The analysis is performed only for major projects and is usually done on a regional basis. While this analysis is a useful and appropriate tool to use at high level (*i.e.,* policy) planning, the social and side effect costs may not be available without some lower level and more detailed environmental impact analysis being performed first. Even if these social and side effect phenomena were determined, the "value problem" remains difficult to solve since many of these phenomena are not easily assigned a dollar cost value. The method is very sensitive to the actual values used (Andersen 1974 and Burnham *et al.*, 1974). Hill and Tzamir (1972) claim that environmental benefit/cost analysis has been a failure. This may be overstating the point. Benefit/cost analysis could be a very effective policy development tool for high-level planning. What is required is to provide as input to such studies the costing results of detailed preliminary assessments of environmental effects.

Land Unit Analysis

Descriptive Land Unit Analysis has been applied for some time in environmental studies. The approach is to assemble a team of specialists who perform a detailed inventory of significant plant communities, wildlife habitats, recreational and cultural sites, established human land use patterns and so forth. They study the terrain topography, and geologic nature isolating major landscape types. An analysis is then performed to associate spatially the inventoried significant communities, land uses etc. with landscape types to derive a set of landscape units. The landscape units, in other words, are subregions which are identified by the association determined between the physical characteristics of the region and the biological-human use conditions.

Vaughn (1973) describes methodologies which can be used to identify land units. When the subregions have been identified and mapped, each is assessed individually with respect to its ability to tolerate, moderately tolerate, or its inability to tolerate the effects of a project. The study on Saltspring Island, British Columbia, by Hirvonen *et al.* (1974) of Environment Canada is a good example. The technique is also used in part by Burnham *et al.* (1974). They include in their analysis a public participation stage which was used to establish evaluation criteria for land unit associations.

The important characteristic of this method is its ability to locate subregions within the study area which are not compatible with the project and other subregions which are compatible. This kind of spatial resolution is not available from checklists, matrix methods or benefit/cost analysis. Indeed, it is conceivable that a benefit/cost analysis could reject a project which would have been acceptable if low-impact (*i.e.*, tolerant) land units had been chosen for the proposed project.

The successful application of this method requires a great deal of careful field work and mapping. This is, perhaps, one explanation of the method's lack of popularity. Like most other methods, descriptive land unit analysis is very dependent upon the professional judgment applied by the members of the study team. The standard method makes heavy use of manual cartographic techniques. Accordingly it is very difficult to make alterations to the study to reflect the emerging requirements of government and the public. Usually, the method does not carry on beyond the basic assessment of the suitability of land units. It is left to the proponents to use other methods to evaluate alternative routings and decide whether to proceed or not with the project. The method does have great advantage over many others since part of its output is in mapped form and can be easily comprehended by specialists and nonspecialists alike.

Input-Output Analysis

Input-output analysis has recently been extended to environmental impact assessment. The basic method was developed in the 1940s and 1950s by economists who wished to study the interdependence of macroeconomic phenomena. For each major industry its set of inputs (*e.g.*, specified raw materials, products consumed) and its set of outputs (*e.g.*, specified final products, unfinished products) are tabulated in matrix form. One dimension of the table specifies inputs, the other specifies outputs. In addition, gross regional demand for various commodities and the government sector are specified. Miernyk (1965) gives a good description of how an input-output matrix is developed. One can show in the table the linking of the outputs of one firm to the inputs of others in addition to the distribution of outputs to final demand sectors. Thus, industry interdependencies can be described in the table. By appropriate matrix operations, coefficients can be calculated which show the total direct and indirect effect in the system when the output of one industry is increased.

The environmental extension of input-output analysis is achieved by including as part of the processing sector, a project, its demands for inputs and its output including environmental contaminant output (possibly broken down into various types). Entries can be included to represent social investments or social costs associated with projects. Nijkamp and Paelinck

(1973) give an example. It is possible to include a number of proposed projects in a study. Input-output analysis can be used in a predictive manner by performing an input-output study with and without the proposed project to see how the solution changes. Coefficients derived in an analysis may be used to estimate the increases or decreases which could be experienced in various pollutants, plant species, agricultural production and so forth. To obtain such predictive power, a matrix must be established identifying all the inputs (*i.e.,* dependencies) and outputs (*i.e.,* influences) of each significant component in some common unit; this is usually the dollar.

In order to take into account the existence of regions with special characteristics which may affect the total solution, the matrix is augmented. Essentially, an input-output model is developed for each region, extra entries are then added establishing the linkages between regions and the input-output subtabulations are combined into one table. Many matrix-ordering variations are possible; numerous papers on model development may be found in the papers of the Regional Science Association and its journal. However, the basic method described above is common to all models.

It is apparent, however, that it is necessary to establish dollar valuations for positive and negative environmental effects along with social costs if one is to completely structure an environmental input-output analysis. Isard and Van Zele (1975) propose urban-form environmental analysis by multiregional input-output. They observe, however, that such detailed analysis awaits the development of adequate data base information. The informational requirements for such studies would be prodigious. Miernyk (1965) reports that an economic input-ouput study of the Colorado River Basin, divided into only six regions and without using substantial aggregation of firms into industry groups required the development of a 300 x 300 matrix of 90,000 entries. In general, N-squared transfer coefficients must be obtained and stored in computer core when N is the sum of the individual inputs and outputs of firms or environmental effects. While there may be some potential for subregional studies, it appears that the quantity of detailed data required will not be readily available. This will limit the input-output approach to aggregate studies on limited numbers of subregions.

Cartographic Overlays

Cartographic overlay analysis techniques have been used to perform detailed environmental assessments. In some cases the technique involves a manual overlay; in other cases, a computer is used to draw maps resulting from an overlay of various map series. This approach has evolved out of the land classification techniques developed by British and American field workers in the 1920s and 1930s. In the land classification problem, one attempts to develop an exhaustive partitioning of space into regions where there is minimal variation between regions of the same class and maximum variation between regions of different class. The initial impetus for land classification systems was to inventory soil type and the value of land for agricultural purposes. The initial idea was to make separate field studies of specific phenomena (*e.g.*, soil texture, soil reaction, soil drainage), map each phenomenon on a separate map series using some simple interval classification scheme, and cartographically overlay the different map series to produce a summary map series. The summary map was visually inspected and generalized into regions with similar character; such assessment series become more interpretive and abstract. Statistical models (*e.g.*, factor analysis and multiple regression analysis) have been used to establish a basis for mapped components. Combinational or decision tables have been used to arrive at partial or final assessments.

Hoffman (1971) provides a good example of a recent application based on statistics. His study was associated with the Canada Land Inventory developed by the Canadian Government under the ARDA project in the late 1960s. The project was heavily based on a series of land classification mapping exercises. Among the various map series produced were: Soil Capability for Agriculture, Capability for Recreation, Capability to Support Wildlife (waterfowl), Capability to Support Wildlife (ungulates) and Land Use. Mapped assessments were manually produced at a working scale of 1:50,000 inches. These surveys were completed for only selected areas of Canada due to their substantial labor component. A major concern in operating a large survey of this nature is to maintain acceptable accuracy from map sheet to map sheet of the same series. The use of different assessors, particularly if judgments are required, gives rise to variability. This necessitates the use of very simple classification rules or algorithms which in turn compromise the precision of a survey and limit its usefulness. However, the method does provide a detailed spatial assessment of phenomena and as observed by Hoffman (1971) has great potential in planning problems.

Alexander and Manheim (1962) were some of the first to apply an extension of basic land classification techniques to route selection. In this case, the application was to route part of Interstate Route 91 in Massachusetts. They identified 26 design requirements and for each requirement performed a land classification. By using grey scale shading (from darkest as most suitable to lightest as least suitable) they developed 26 maps which were then overlayed to provide a summary assessment of areas suitable for highway development. To increase the effectiveness of the method, the structural nature of the 26 requirements was analyzed using "hierarchial set decomposition" (a form of cluster analysis) to find groupings of similarly structured requirements. They then combined two maps at a time in similar groups until there was a subcomposite map for each nonsimilarly structured group. As pairs of maps were combined, their relative shading weight was adjusted photographically to obtain the maximum pattern definition. The results of this process were manually redrafted to be used in the next overlay stage. A final map was obtained in a similar way by combining the subcomposite maps. This final composite map was examined visually to see if a dark (or highly suitable) route could be located. No feature of their methodology guarantees continuity of a route. Indeed an important value of this form of analysis is that the routing proper is left until the basic suitability analysis has been completed. From this kind of process, it is very possible that no route may be found which is "suitable" over its entire length.

McHarg (1971) independently developed a similar graphical approach. He has extended its application from an initial study on locating an interstate highway in New Jersey through analyzing the development potential of Staten Island in New York City to analyzing development potential of metropolitan regions. Similar to Alexander and Manheim (1962), McHarg identifies numerous factors related to the problem, maps them and combines the maps into a composite. His factors tend to be interpretations based upon physical or natural phenomena and existing land uses. Unlike Alexander and Manheim, he uses color in addition to shade intensity. This gives him the possibility of carrying both intensity and compatability assessment through the cartographic process. Both the McHarg and the Alexander and Manheim studies make use of intensity or color adjustments in the map combination process. This effectively alters study component weights in an unspecific way. While it provides the reviewer with a composite assessment it is not readily clear how the combination has taken place.

Information System Analysis

Computer-based information systems have recently been applied to environmental assessment problems. Turner (1970, 1972) credits P. O. Roberts with originating in 1957 the idea of combining photogrammetric and air photo interpretations with computer data storage, manipulation and mapping. However, one of the earliest large-scale studies utilizing a computer-based information system was performed by Neiman and Miller (undated) at the University of Wisconsin in the early 1960s. Their work could be judged the transition from the exclusively manual cartographic techniques of McHarg, Alexander and Manheim to studies which involve significant computer analysis. In the Neiman and Miller study, the study area was broken into arbitrary 1-km units or cells, map series were analyzed and a classification record was made in an information system for each 1-km unit. A number of such variables were inventoried. The computer was used in a bookkeeping mode simply to store the data and draw maps on a cell by cell basis using a standard printer mapping program.

Denenberg *et al.* (undated) have recently extended the idea of a computer-based environmental information system. They have developed NARIS (National Resource Information System) at the University of Illinois to provide, on an interactive basis, access to a regional data bank of terrain information. A region of Illinois has been arbitrarily divided into 40-acre units for each of which a number of data variables are recorded. The user can type requests formulating simple models and have the results produced in printer map form. The NARIS system shows great promise for the investigation of linear weighted environmental models and for instructional purposes.

Krauskopf and Bunde (1972) at the University of Wisconsin have developed the initial work by Nieman and Miller by refining data base resolution of phenomena recorded. They use an inventory unit which is 500 meters square and keyed to the standard Universal Transverse Mercator (UTM) grid used in the topographical survey. Their work identifies a number of data variables which are useful in environmental studies. They propose the separate evaluation of "factors" and their subsequent combination by linear weighted models. Rattray *et al.* (1973) also use simple linear models. They observe that this is still essentially a bookkeeping use of the computer and propose that more complex assessment models utilizing logical models and threshold analysis might be appropriate.

Robinette (1973) and Lyle and Von Wodtke (1974) report on applications similar in nature to the work by Krauskopf and Bunde. Both applications use linear weighted models applied to a regional data bank to produce factor and composite maps. Robinette does not given descriptions of data elements required for such a study. Data base development and the actual models are left undefined. In addition, the increased power of a computerized information system is not used to simplify the presentation of analysis results. In his study of siting power-generating facilities 91 different maps are produced, and while some are composite "interpretations" of others there is no single composite assessment map. Lyle and Von Wodtke have applied the technique to a small regional development planning problem. This application is over an 82-square mile area with basic cell size of 111 feet or 0.28 acres. When appropriate, data are aggregated to 1000-foot square units with areas of 22.9 acres. This indicates the potential usefulness of the basic grid analysis technique on small-scale planning problems.

In general, the information system studies appear to apply simple linear weighted models to regional data banks. Little is described regarding data collection and data base flexibility. It appears that it is relatively difficult to modify the data base while studies are progressing. To date, there has been little advance beyond the fundamental land classification approach.

Analysis by Mathematical Surfaces

Turner (1970, 1972) and Hausmanis (1971) have used mathematical surface approximations for environmental assessment of highway construction and have included a routing phase. The computer software system used is called GCARS (Generalized Computer-Aided Route Selection), and is currently being used by the Ontario Ministry of Transport and Communications. The system is based on surface analysis. A number of data points are fixed in space by geographic coordinates (X,Y); the data value at a point is referred to as Z_i for each variable i. Thus, arbitrary point j in space may have the information record

$$X_j, Y_j, Z_{i,j}, \ldots, Z_{i,j}, \ldots, Z_{m,j}$$

for some m number of variables being recorded. If one considers each arbitrary Z_i as belonging to the surface described by the Z_i values distributed over the X,Y space, then a functional relationship

$$Z_i = f(X,Y)$$

can be defined where the function f is a mapping of X,Y to Z_i at some arbitrary order of f. Various mathematical techniques are available to estimate the coefficients of such functions. However, all are based on the concept that values of Z_i are mathematically continuous. The advantage of this technique is that only a sample of points is required to estimate the coefficients of f. Once obtained, the coefficients of f can be used to estimate a $Z_{j,k}$ value for any arbitrary point (X_k, Y_k) by simply defining:

$$Z_{jk} = f(X_k, Y_k)$$

The accuracy of such an estimate for $Z_{j,k}$ is dependent upon the coefficients estimated for f. This in turn is dependent upon the selection of an appropriate degree relationship for f, the selection of appropriate data points to define the surface (both number and spatial distribution) and availability of reliable estimating programs for the function degree selection. Each function f describes a surface.

Turner and Hausmanis use a function g which is a combination of weighted f functions on different raw data variables to provide an estimate of a "highway location factor." In addition, they use a weighted combination of the g functions (or factor surfaces) to define a composite function or surface. This composite surface is contoured and also used as an estimate for a utility surface measuring suitability for the development of a highway route.

An arbitrary grid is established over the utility surface and utility values are interpolated for grid intersections. These utility values are used via a minimum path algorithm to select routes. While Turner acknowledges the existence of discontinuous and discrete measures, his system assumes continuous variables. For example, the four grid interpolation techniques available in GCARS all assume a continuous metric. Turner (1972) notes the difficulty in determining the optimum sampling rate to obtain accurate surface estimates. The same weighting problems encountered in the overlay and information system approaches are observed in Turner's system. The use of linear weighted surface averaging techniques appears to impact on the resolution in the route-finding stage. The examples Turner gives show a tendency for straight-line routes.

Other Methods

Various other methods for spatial planning are being developed. Many are variations of the nine techniques described above. Techniques of combinatorics and operations research have been proposed as ways of making sensible spatial resource allocation decisions. These are surveyed in Scott (1971). Few have been successfully applied in environmental impact studies. Typical of these methods is the work by Gordon and MacReynolds (1973) where they discuss the problem of optimal assignment of a set of activities to a set of regions. Fundamental to their method and most of those described by Scott is the assumption that relative locational advantage can be determined by dollar value cost or profit matrices. These methods share many of the same difficulties as benefit/cost or input-output analysis. Few propose routing methods. However, some useful stage-wise planning models are available after basic routes are determined.

METHODS FOR ROUTE DETERMINATION

Methods for route determination have been of interest in mathematics, operations research, and computer science since the 1950s. Most methods structure the problem as finding a minimum-cost path through a network or graph. In one variant, cost is equated with cartesian or straight-line distance. The major impetus for the development of route-finding algorithms came from the electrical engineering problems of routing telephone traffic, designing wiring plans for computing equipment and the circuit layout of printed circuit boards. In addition, path-finding techniques find application in mechanical theorem proving and problem solving. While none of these methods were developed in the context of environmental routing problems, some have found application there (Turner 1970, 1972). Recently, one heuristic based on an established mathematical method has been developed in the context of an environmental routing problem. To facilitate discussion of the various methods available and the later discussion of new methods developed, it is necessary to give a formal definition of some elementary concepts in graph theory. Extensive discussion of these and related concepts may be found in most texts on graph theory. Stigall and Tasar (1974) provide a readable survey. Knuth (1969, 1973) and Berztiss (1971) provide definitive discussion of basic graph theory. Each of these authors discusses methods of storing and manipulating graphs in computer programs. Hart *et al.* (1968) provide a formal discussion of the general framework for determining minimum-cost paths.

Some Graph Theory Definitions

A graph is a mathematical model of a system. It exhibits a relation or the absence of a relation among the elements of a special set V. The members of this set are "vertices," sometimes called "points" or "nodes."

Definition 2.1: A *graph* G is an ordered pair $G = \langle V,R \rangle$ where V is a set of vertices and R is a relation in V represented as a set of pairs of elements of V.

Definition 2.2: If $x,y \in V$, then x is said to be *adjacent* or *connected* to y if the pair $(x,y) \in R$.

Definition 2.3: An *edge* is a pair $(x,y) \in R$ with $x,y \in V$. Sometimes an edge is called an arc or line.

Definition 2.4: The *edge set* E of a graph $G = \langle V,R \rangle$ is the set of all ordered pairs $(x,y) \epsilon R$ for all $x,y \epsilon V$.

It is usual to refer to a graph as $G = \langle V,E \rangle$ where E is the edge set resulting from the relation R in V.

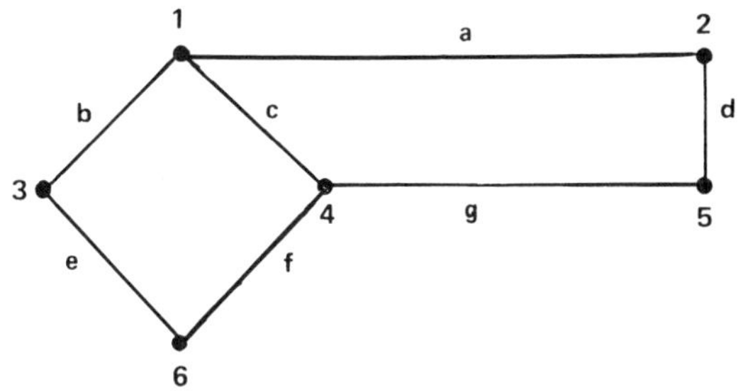

Figure 2.1. A sample graph.

An example of a graph may be seen in Figure 2.1. This Graph $G = \langle V,E \rangle$ is defined by the following sets:

$$V = \{1,2,3,4,5,6\}$$

$$E = \{(1,2), (1,3), (1,4), (2,5), (3,6), (4,6), (4,5)\}$$

The elements of E as enumerated correspond to a,b,c,d,e,f,g in the diagram.

Definition 2.5: If $x,y \epsilon V$, a *directed edge* (arc, line) is an ordered pair (x,y) of the adjacent successor relation R where equivalently y follows x or y is the successor of x or x points to y. The introduction of directed edges required R to be an ordered pair relation and permits the concept of direction to be added to a graph.

Definition 2.6: A *digraph* $G = \langle V,E \rangle$ is a graph where E consists of directed edges. The previous example graph has had direction added as indicated by the arrows in Figure 2.2.

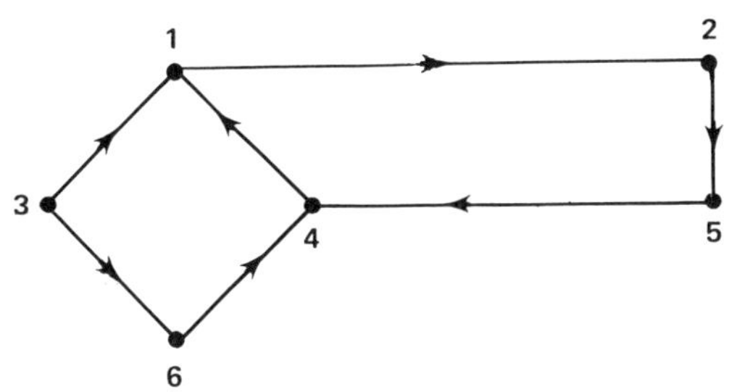

Figure 2.2. An example of a diagraph.

This *digraph* G = < V,E > is defined by the following sets:

$$V = \{1,2,3,4,5,6\}$$

$$E = \{(1,2), (3,1), (4,1), (2,5), (3,6), (6,4), (4,5)\}$$

Definition 2.7: If G = < V,E > is a digraph, the *invalence* of some x ∈ V is the number of edges in E where x is a successor. The *outvalence* of x is the number of edges in E of the form (x,y). For example, in the previous digraph, the outvalence of vertex 5 is 0 and its invalence is 2.

Definition 2.8: An *out-directed tree* is a digraph G = < V,E > when there exists a unique x ∈ V, called the *root*, which has invalence of zero and outvalence greater than zero. All other y ∈ V have invalence of exactly one. There is a set L with L ⊂ V such that each y ∈ L has outvalence zero; these elements are called the *leaves* of the tree.

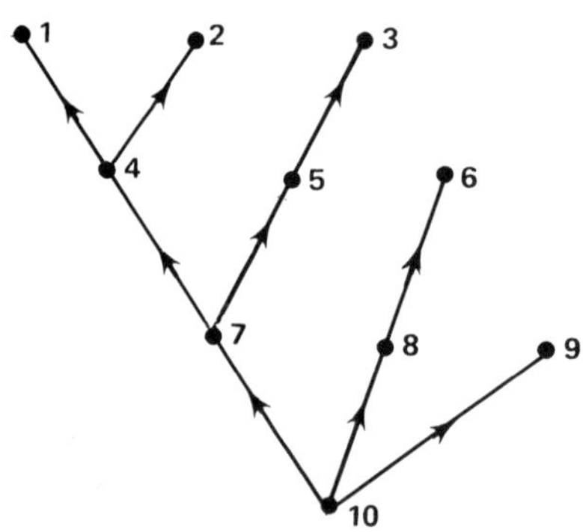

Figure 2.3. An out-directed tree.

Figure 2.3 gives an example of a tree with root at vertex 10 and vertices 1,2,3,6,9 as leaves.

Definition 2.9: A *path* from $x \in V$ to $y \in V$ in a digraph G = < V,E > is a sequence of edges in E of the form

$$(x,P_1)\,(P_1,P_2), \ldots, (P_i, P_{i+1}), \ldots, (P_{n-1},P_n), (P_n,y)$$

where each successive ordered pair has as its first element the successor of the previous ordered pair.

In Figure 2.3 various paths may be detected. Some examples are the path from the root to the leaf at vertex 2 which is (10,7), (7,4), (4,2), or the path from vertex 7 to the leaf at vertex 3 which is (7,5), (5,3). One should note that there is no path from vertex 1 to vertex 10.

Definition 2.10: A *weighted graph* G is the graph $G = \langle V,E,W \rangle$ where W is a set of weights or values or costs associated with each edge in E.

Definition 2.11: A *polygon* (or a *circuit*) is a least path of length greater than 2 from some vertex X back to vertex X with two distinct edges adjacent to X.

Definition 2.12: If $X \subset V$ for some graph $G = \langle V,E \rangle$ the *subgraph generated* by X and denoted G[X] consists of the vertex set X and the set $A \subset E$ such that each end of an edge in A is a vertex in X.

Definition 2.13: If $X \subset V$ for some graph $G = \langle V,E \rangle$ the *coboundary* of G[X] is the set $B \subset E$ such that for each $b \in B$ one end is a vertex in X and the other is a vertex in V-X.

Definition 2.14: In a digraph, the coboundary of G[X] is out (in) directed if $\forall\, b \in B$ are directed out of (into) of G[X]. A coboundary is directed if it is either in directed or out directed.

Definition 2.15: A graph $G = \langle V,E \rangle$ is *strongly connected* if the coboundary of each non null $X \subset V$ is not directed.

Definition 2.16: A *strong component* of a graph G is a maximum non null strongly connected subgraph of G.

Mathematical Methods for Route Determination

The Transhipment Model

The Transhipment Model is an extension of the transportation problem which is itself a special case of linear programming. A specialized matrix technique easily implemented on a computer is outlined in Cooper and Steinberg (1970) to solve the transportation problem. The method itself was developed to determine the least-cost allocation of some product between a set of suppliers and consumers. The solution is based on a complete matrix of transportation costs between source and destination. While this method is not directly helpful in the routing problem, its extention in the form of the transhipment problem has been used for small systems. The transhipment model is a transportation model with the addition of a set of transfer or transhipment points. These points are added to the transportation model simultaneously as sources and destinations with an arbitrarily large supply/demand of product. The transportation cost matrix is augmented by the addition of inter–transhipment point costs and the costs between transhipment points and the sources and destinations. The resulting matrix is then solved as a standard transportation problem.

A shortest path problem can be solved by this method if all vertices of the graph through which the path is to be developed are considered as transhipment points. The source and destination are specified with infinite distance between them and the matrix of inter vertex distances is developed. When N is the number of vertices in a graph, this structuring of the problem requires the storage of a N+1 by N+1 matrix. Clearly, this technique is not tractable for even moderate sized graphs due to its excessive storage demands since storage needs grow on the order of the square of N.

A Linear Programming Solution

The basic form of the linear programming problem is to minimize (maximize) a linear function (called objective function):

$$Z = C_1 X_1 + C_2 X_2 + \ldots + C_k X_k$$

over all those points in the region where $X_i \geqslant 0$ for $i = 1, \ldots, k$ which also satisfy a set of m linear inequalities (or constraints)

$$A_1 X_1 + A_{12} X_2 + \ldots + A_{lk} X_k \leqslant B_1$$

$$A_{21} X_1 + A_{22} X_2 + \ldots + A_{2k} X_k \leqslant B_2$$

$$\vdots$$

$$A_{m1} X_1 + A_{m2} X_2 + \ldots + A_{mk} X_k \leqslant B_m$$

The values for variables X_i for $i = 1,k$ are predetermined. These are called the decision variables, Cooper and Steinberg (1970) describe how such a system can be organized in a "simplex tableau" and solved. As shown by Scott (1971), to use the method for route finding, inter vertex distances are specified as the coefficients of the objective function. This requires an objective function with the square of N terms when N is the number of vertices in the graph. In addition, the constraint equations must be specified in a matrix of dimension 2(N-1) by (N-1) (N-1). Efficient programs are available which require time on the order of 1.5 M when M is the number of constraints. However, their storage requirements increase as the order of the cube of N, the number of vertices. Clearly, the prodigious memory demands of this method make it useful for only very small problems.

The Dantzig Algorithm

While Dantzig developed the statement of the shortest path problem in terms of the linear programming model (Scott 1971), he is also well known for the development in 1960 (independently from Dijkstra) of the routing algorithm which bears his name. Scott describes an application of this algorithm. The algorithm iteratively develops a shortest path directed tree from the source to all vertices in the graph. The cost of any edge chosen for the path is the weight or cost of the edge. Accordingly, it is necessary to maintain a table of weights corresponding to the edges. However, unlike the two methods described previously, it is not necessary to store distances for all possible edges whether or not they are in the graph. This achieves a substantial reduction in memory demands. A detailed description and example may be found in Berztiss (1971).

The algorithm starts at the source vertex; it assigns the source vertex to a set which will become the set of vertices to which it has calculated a shortest path. Iteratively edges are inspected to find all those (x,y) where the successor vertex y does not belong to the short path set but the other vertex x does. The algorithm maintains for each vertex in the short path set the length of the shortest path to that vertex from the source. At each iteration, the cost of (*i.e.,* length) of each existing path plus the length of the path when the next edge is added is examined for each vertex in the short path set. The least such extended path cost edge is selected for path development. As each edge is selected, it is added to the directed tree. The algorithm stops when no further edges can be selected. The tree which has been developed can be scanned for the destination vertex and traced back to the source to determine the shortest path. This requires an appropriate double-threaded list structure (Knuth, 1969) to be developed and maintained to store the tree.

The algorithm is general; weights need not be symmetric (*i.e.,* the distance from x to y does not have to equal the distance from y to x). However, the algorithm is costly since it develops all possible short paths before termination. Berztiss (1971) describes a double application of Dantzig's algorithm simultaneously from the source forward and backward from the destination. The two associated short path sets are examined for intersection on an identical vertex and then the short path is traced. The implementation of this double algorithm is complex, but performance is improved over the simple Dantzig algorithm. As reported by Pohl (1969) the "best" implementation of Dantzig's algorithm requires time on the order of the square of N when N is the number of vertices in the graph.

The Dijkstra Algorithm

Dijkstra's (1959) algorithm also is based on the concept of developing a directed shortest path tree outward from the source vertex "s." Pohl (1969) credits Dijkstra with the initial development of the tree strategy. Dijkstra defines three sets A, B, and C as follows:

> A–the vertices having their minimum path from the initial or starting vertex;
> B–the direct successors of the above set A which are not in it; and
> C–the remaining nodes.

The algorithm proceeds in two stages:

1. A vertex in set B with current minimum distance (*i.e.,* weight or cost) to s, the source, is transferred to set A. If the vertex is t (terminus) the path has been determined and is retraced from the tree.

2. The successors of the vertex just placed in A by step 1 are calculated. Of these, the nodes that are in C are transferred to B. The value of the distance from s, if the vertices already in B are changed is the new distance calculated is smaller than their current value. Repeat step 1.

Dijkstra's and Dantzog's methods are very similar. Pohl (1969) reports that Dijkstra's basic algorithm requires time order of the cube of N, the number of vertices. Yen (1972), Pohl, and Williams and White (1973) report recent improvements; however, the best implementation requires time on the order of the square of N.

The Lee Path Connection Algorithm

C. Y. Lee (1961) developed an algorithm in the context of implementing wiring diagrams. He visualizes the area through which the path is to be routed as being divided into a large number of identically sized cells. The cells can be used to store information pertaining to path development. Arbitrary data values are stored in the source cell, destination, and forbidden cells (*e.g.,* 999, 9999, 99999) so that these areas can be detected.

The algorithm works outward from the source by checking immediate neighbor cells (to the north, south, east or west) in iterative steps. It marks its progress by storing the iteration step value in each cell not already having a value. The value "1" is stored immediately around the source. It then attempts to store as many "2's" as possible around the 1's. This process is continued storing as many "3's" around "2 s" and so forth. Eventually all cells will be filled or the cell containing the destination is located. In the former case, there is no path between source and destination; in the latter, a path exists of length equal to the current iteration of the algorithm. A path can be retraced easily by using the successive iteration step values stored in the matrix to backtrack.

A major limitation of this algorithm comes from its reliance upon the distance between any two neighbor cells remaining constant. This makes it impossible to include directly in route determinations, real costs, or weights which may be nonlinear or discontinuous throughout the study area. In addition, the study area is polarized into only two classes of subregions: areas completely suitable and areas entirely unsuitable for path development. No provision is made for classifying areas for intermediate suitability; indeed, it is not apparent that this could be done in any general way based on Lee's method.

The analysis grid size must be set to the maximum resolution required. Akers (1967) and Aramaki *et al.* (1971) have developed storage saving techniques but they claim the practical limits of Lee's algorithm when applied to automating etching-pattern layout is roughly 37,000 cells. This provides resolution of 1.27 mm on a circuit board of maximum size 20 x 30 cm. With the advent of microminiature devices and more complex circuit boards, greater resolution is required. For example, reducing to 1.0 mm accuracy from 1.27 requires a Lee path analysis of 60,000 cells. When N is the square root of M, the number of cells, the time requirement for Lee's algorithm is of the order of the cube of N. To date the algorithm has received wide use in circuit board layout problems in spite of its memory and processing demands. This is in large part due to the ease by which it may be implemented on a computer.

Heuristic Methods of Route Determination

Various *ad hoc* methods have been developed for locating paths. Most have been based on or are extensions to established mathematical algorithms. The heuristic methods by Aramaki *et al.* and Goodchild and Potts (1975) have had some success in applications. Both methods are based on gridded study areas.

The Aramaki-Kawabata-Kazuhiko Heuristic

Aramaki *et al.* (1971) developed this method to overcome the computational cost of Lee's algorithm. They first attempt to find paths heuristically and, in event of failure, they apply a modified version of the Lee algorithm. Their method is based on a straight-line point-to-point connection with permissible directions for paths being via 90-degree turns and straight lines. The heuristic proceeds in the following three steps:

1. Draw horizontal lines from the source and the destination as far as possible in the general direction of the other.
2. Find a vertical path within the rectangle bounded by the vertical separation of the two straight lines and their horizontal "overlap."
3. If such a vertical path cannot be found in the second step, assume tentative target points up to plus or minus two cell widths removed vertically from the original points. Apply the first and second steps with the tentative points.

The algorithm essentially constrains routing to that subregion of the study area which is the least rectangle containing the source and destination points. In addition, it does not permit the orderly development of paths which could go around an obstacle by routing a subsection of the route away from both source and destination for a short time. The authors note that while this method is fast and usually successful, it is often necessary to fall back upon Lee's algorithm when "difficult" cases are encountered. Since the heuristic is based on a gridded study region approach, its memory requirements are approximately the same as the Lee path algorithm.

The Goodchild-Potts Heuristic

Potts (1975) reports the recent development in cooperation with Goodchild, of a heuristic for route finding. The algorithm is applied to a gridded representation of the study area. Unlike the Lee algorithm it permits intercell weights to be nonuniform. The method is based on using aggregate region analysis to select a possible route and subsequently refining the route in successive iterations. The user specifies an iteration limit; this in turn determines the maximum region size for iteration one. Square subregions are developed replacing a number of basic analysis cells. Each of the cells in a subregion may have a number of cells of quite different weight. The heuristic assigns a single weight to the aggregate subregion. This is achieved by developing a frequency count of each cell weight detected. The aggregated subregion is assigned the highest frequency weight. The developers of the technique prefer this to calculating an average weight since "the averaging procedure could be influenced by very localized pockets of [extreme] high or low values within the aggregated cell." Accordingly the study area is broken into weighted aggregate subregions through which a route is developed using a modified Dijkstra algorithm. The distances or weights between all linkages are assigned as the average weight of the two cells. By applying the modified shortest path algorithm, a list of aggregate subregions is determined through which the route will go. Further development of the route in subsequent iterations is entirely within this set of subregions.

For second and subsequent iterations, the algorithm is reapplied each time based on a study area consisting of the set of aggregate subregions selected at the previous iteration. Aggregation is performed to a one less aggregation level with each step resulting in analysis at the individual cell level at the last iteration. By adopting this iterative approach with precision increasing with each step, substantial savings are made in storage requirements and correspondingly larger studies can be accommodated.

While this heuristic will find a path, there is no guarantee that the path will be the shortest or least-cost one. This is due to the aggregating together of unlike cells and assigning to the aggregate some arbitrary weight. The maximum frequency approach the authors selected can give an erroneous generalization to a region. As an example, consider aggregating a 4 x 4 region where weights had been assigned on a ten-point scale. Suppose one found the combination of: four 1's, three 7's, three 8's, three 9's and three 10's. The maximum frequency weighting would assign the weight of 1 to the square but 1's account for less than 26% of the regions. If the weights had corresponded to impact assessments where 10 indicated prohibitive impact and 1 indicated no impact, the maximum frequency weighting hides the fact that heavy to prohibitive impacts dominate the region. The effect on route determination can be catastrophic. The use of the weighted average would have given a score of 6.6. Another weighting for this cell might have been 3.5 which is the weighted average of the high scores accounting for 75% of the area.

A further problem is related to this generalized or aggregating technique. Figure 2.4 shows the possible configuration of four 3 x 3 aggregate cells and the individual cell weights. If it were necessary to develop a path from bottom to top, it is clear that the path should go through cell **D** where a string of 1's are available. Both maximum frequency and weighted average assign a lower weight to cell **B** than cell **D**. The nature of the heuristic would cause cell **B** to be accepted for *all* future path development. However, the frontier of 10s in cell **B** will cause a very expensive route to be developed in a later stage. Where the weight 10 should be considered "no development" or "no go" we wish to prohibit paths from going through any cell weighted 10. In such a case, the heuristic would have selected as a development subregion an area where no route was possible; yet, outside the subregion, a path was available. The major difficulty with this algorithm is associated with the effects of generalizing or aggregating unlike cells to form subregions. This was done to assist in removing constraints on problem size. The result is a partial constraint on the accuracy of the resulting analysis.

	Cell "A"			Cell "B"			Cell "C"			Cell "D"		
	8	10	10	10	10	10	10	10	10	1	10	10
	10	10	6	2	1	5	5	10	10	1	6	10
	10	6	2	1	1	1	3	5	9	1	4	10
Maximum Frequency	10			1			10			10		
Average	4.5			2.6			4.5			3.3		

Figure 2.4. Sample aggregate cell configuration.

ENVIRONMENTAL ASSESSMENT: RELATED PROBLEMS

In the preceding review, problems related to specific environmental assessment methodologies were outlined. In addition to those problems a number of major but general problems may be identified.

The Value-Ranking Problem

This is perhaps the most difficult of all problems in environmental impact assessment. Fisher and Davies (1973), Orloff (1972) and Dickert and Sorensen (1974) observe that the fundamental problem is lack of agreement on appropriate methodology to use and effects to consider. In addition for nearly every methodology some choice must be made concerning the parameters to analyze. While the U.S. Environmental Protection Agency (1973), Leopold *et al.* (1971) of the U.S. Geological Survey, Resources of the Future (1972) and others have proposed effects to be considered for various classes of projects, many projects have not been included in these lists. Dickert (1974) observes that indirect effects are rarely included. As a result, the group doing an analysis must choose the specific effects to consider. In addition, once the effects have been selected, the measurement and scaling of the effects require further judgment. Often studies are criticized for not making specific the judgments taken in the study.

Reveiwers (Canadian Environmental Law Associaton, 1973; Lucas, 1973) comment that many methodologies do not encourage making the judgments explicit.

Most of the quantitative techniques require judgments in scaling effects in common dollar value units. In many cases this involves a difficult judgment. Andersen (1974) wonders "why should 'willingness to pay' be the monitor of human desires and needs." Bishop (1974) and Klassen (1974) observe how very sensitive some methods are to the dollar values assigned. Jowett (1972) remarks "aesthetic judgments are at best ordinal." Current methodology does not appear to have assisted greatly in reducing the value judgment problem. Indeed some have made it worse by forcing dollar values to be placed on intangible items. It appears that some assistance could come from a quantitative method which could accommodate simultaneous use of dimensionless variables and assessment rules as well as the more usual continuous variables and models.

The *No Development* Problem

Closely allied with the Value-Ranking Problem is the *No Development* or *No-Go* problem. Only the Benefit/Cost approach appears to include a specific consideration of whether or not to go ahead with the project. It is, however, this question which is most fundamental for a project. No approval for project development is one possible outcome of the project review process. Usher (1973) observes that this question is usually ignored or not part of assessments. He observes that to be effective, a methodology must explicitly accommodate the possibility of no development. Malisz (1972) also argues for definition of the No Development option.

When routing a utility, it is obvious that certain subareas of the study area must be considered *no go* or *no development*. This could be true even for construction reasons alone. For example, it does not make sense to route a highway through a large lake if an alternate land route is available. Various environmental effects could also establish a subarea for No Development. To be effective, an assessment and routing methodology for utilities must permit evaluation of the No Development problem at a subregion level. This would permit a project level assessment of the No Development problem. If no routes can be located which avoid No Development areas, the No Development status of the total project is determined.

Study Comprehendibility and Public Involvement

Most environmental assessments are destined for public review under governmental requirements. Lucas (1973) notes that this requires the results of the analysis to be easily comprehended and the inclusion of public concerns and perceptions in the study. The Canadian Environmental Law Association (1973), Orloff (1972), and Burnham *et al.* (1974) emphasize the necessity of including public involvement in any environmental assessment. Most assessment techniques do not provide for accommodation of public concerns and perceptions. Input-output analysis is likely the worst offender with its massive tableau of technical coefficients. Benefit/cost analysis is not much better with its use of discounted weighted dollar value assessments of intangible parameters. The cartographic overlay technique appears to provide the most comprehensible results especially if supporting analysis maps are provided. The actual direct use of data reflecting public concerns is rarely encountered.

Data and Analysis Discontinuity over Space

Malisz (1972) observes that data discontinuities are very common in regional planning problems and must be accommodated in methodologies. Isard and Van Zele (1975) observe that detailed subregional analysis is required for this reason. A small area of only a few acres in extent could contain environmentally sensitive phenomena while an adjacent area is free of all impact. Clearly, an assessment methodology for utility routing must accommodate and analyze such discontinuities.

Analysis Scale and Planning Requirements

The nature of each environmental assessment methodology establishes a scale or level at which its application is most appropriate. Many studies involving checklists, matrix methods, benefit/cost analysis and input-output analysis are high-level assessments across a study area. They might conveniently be called macroanalysis methods. Mathematical surface approximation techniques make use of trend surface fitting and smooth data. In other words, the technique generalizes and is most appropriate for higher-scale studies. Only the cartographic overlay and computer-based information systems appear to be fully capable of detailed low-level evaluation.

The environmental effects of a utility route are on a spot-specific basis. It is essential that a detailed localized study be performed if a realistic environmental impact assessment is to be made. The implication is that a proper high level of policy level assessment of environmental impact cannot be made without an adequate low-level assessment first being performed to provide an accurate environmental cost estimate. This point of view is shared by Andersen (1974), Twiss (1974), McNeil (1973) and Fisher and Davies (1973).

While the cartographic methods and computer-based information systems show promise for the detailed low-level analysis, the scale of important physical phenomena gives rise to difficulty. Steiner and Stanhope (1972) show that 40-acre analysis resolution is necessary for regional planning applications. This requires analysis accuracy in the order of 0.25 inches on a 1:50,000 topographic map sheet approximately 16 x 22 inches in size. Data acquisition and record keeping for the 2500 units per map sheet is a difficult task requiring the use of a computer.

Study Flexibility

The ongoing development of environmental standards and increased public involvement in assessments gives rise to adjustments being required to study methodology either in the analysis or the review phase. This requires an analysis method which can relatively easily accommodate added (or deleted) data as well as changes in the valuation of weighting of the analysis model. Few of the established methods contain much of this flexibility.

Usher (1973), The Ontario Ministry of Environment (1973), McNeil (1973) and the Canadian Environmental Law Association (1973) note the necessity for proper examination of alternatives. To be fully successful, a methodology must not only provide assessment of alternatives but also be capable during the study phase of assessing alternatives proposed by intervenors during the review stage. This calls for added flexibility in the methodology.

GENERAL LIMITATIONS OF ROUTING METHODOLOGIES

In previous discussion limitations of specific path-finding techniques have been outlined. Some more general limitations and problems are now addressed. Routing a utility across even a moderate study area and maintaining adequate routing accuracy requires consideration of tens of thousands of nodes in a network. Most implementations are more or less constrained in the problem size which can be handled. As reported earlier, even the best implementations of the Dijkstra Algorithm required computer time of the order N squared when N is the number of nodes in the route. A method is needed to reduce the complexity of route finding and simultaneously maintain adequate accuracy.

All methods outlined produce routes which are based on some grid over the study area. Resulting routes are irregular as they progress from grid square center to center. Up to now, the only way to smooth out routes has been to reduce the size of the grid. However, as the grid size is decreased, the number of cells increases by the square of X when X is the number of cells replacing a single original cell. Accordingly, N, the number of vertices can be rapidly increased and easily pushed beyond a computationally tractable number. A routing method is required which will produce smoothed routes without reducing problem size.

Other than Lee's method, most algorithms force the development of path from source to destination. When considering environmental "costs," route-finding algorithms must accommodate the *No Development* or *No Go* assessment, and, in addition, accommodate intermediate impact rankings. The generation of alternative routings (even with slightly increased routing costs) is required but is not included as part of the standard methods. In addition, utilities have spatial dimension and cannot be considered as simple lines. This requires consideration of minimum width routings in the algorithm. None of the methods provide this capability.

The established routing algorithms, with the exception of the Goodchild-Potts (1975) heuristic, consider routings by distance only and report only path length statistics. However, when routing a utility across a study area, the path will cross various classes of impacts. It is important that routing methods be extended to provide a detailed report of the nature of each path and its alternates.

CHAPTER 3

DEVELOPMENT OF PLANNING SYSTEM GOALS

The general planning process involves complex interaction of proponents, governments, professional assessors, intervenors and the public. It is proposed that a planning system be developed which will be able to react flexibly to the needs of such processes and accommodate routing problems for all utilities. It is not possible to dictate a specific structure to be followed for the total planning process. Rather, a system is required which will provide an orderly analysis framework within the general process. The system should be oriented to developing relatively narrow alternative corridors wherein a utility could be located subject to a wide and varied set of conditions. Each corridor would be described in detail and its prospective impact quantified to facilitate policy decisions regarding project development. Of course, the final decision is usually made by government and may well be a political one based both on the results of a study and the public reaction to it. Detailed environmental costs derived will provide much of the missing information which has caused difficulty in applying benefit/cost analysis at the policy level.

The lack of direct applicability of existing environmental assessment and routing methods to utility routing has established the need for a planning system which will develop specific routes while avoiding the deficiencies outlined in Chapter 2. The system will be applied by professional assessors. It must allow for the input of public and governmental concerns, knowledge and attitudes throughout its course. In particular, it should facilitate a meaningful public review and be flexible enough to permit the required adjustment resulting from the review. It follows that the questions of scale, study size and discontinuities, system modularity, data base development, analysis techniques and routing methods are central to an effective, flexible planning system design.

SCALE, STUDY SIZE AND DISCONTINUITIES

The planning system should be general and flexible enough to permit its basic methodology to be applied on any scale project. This would be accomplished by changing the scale of the data base and analysis procedures. The current concern over routing utilities through extensive tracts of land requires that the system's emphasis be toward moderate to large-scale projects. While the method could be applied to small-scale projects, it is anticipated that they would continue to be located and assessed for environmental impact by direct field work. The planning system should be used to reduce the large-scale problem to a tractable "solution space." It is not expected that the planning system will perform precise alignments but rather select a series of narrow "corridors" in which precise alignments can be determined. This strategy is required since standard data sources are not available at sufficient resolution to permit such detailed alignment studies. The final alignment or "center lining" stage must continue to be accomplished with the aid of detailed field surveys.

The suitability of corridors for eventual location of precise alignments requires that the analysis be sufficiently accurate to detect and process any discontinuities in phenomena which could affect planning at this scale. As shown by Steiner (1972), Peuker and Chrisman (undated), and Goodchild (1973a,b), and implemented by Deneberg *et al.* (undated) and Krauskopf and Bunde (1972), this necessitates detailed assessment of the study area on an arbitrary grid square basis. Grid cell size will be set small enough to accommodate the significant data discontinuities and should be related to the level of resolution available in data sources. The planning system should be flexible enough to work at various scales over study areas from 1,000 to at least 100,000 square miles in extent.

The planning system must serve all significant requirements for analysis. This will include data acquisition and storage, development of analysis modules, their evaluation and composition of module assessments, route finding, generation of alternatives and their assessments. The system should be structured to permit revisions to be made to model development routing alternatives on an iterative basis. As observed by Malisz (1972) and Rattray *et al.* (1973), the volumes of data to be obtained, stored and analyzed necessitate the use of a computer to implement this system.

PLANNING SYSTEM MODULARITY

A number of distinct system phases may be determined along with a set of tasks to be performed during a project analysis. Study phases produce information which is used as input by subsequent phases. As a study advances, one observes a series of transformations of basic information (*i.e.,* the data base) according to arbitrary models and parameters. This has as an analogue the process stream in a chemical refinery where a series of distinct transformations or unit processes are applied to certain inputs. Johnson (1972) has developed a modular approach for chemical plant design and optimization. A number of individual processes are defined. For each of these a module is developed which performs the appropriate transformations on product streams. These streams serve as a set of inputs and are produced as a set of outputs by unit processes. Each of these processes is developed independently from the others. They can be added or modified as required without requiring modification to other processes. It is possible to develop a unit process library for use in future analysis applications. The planning system analysis and routing phases should be modularized in a similar way. This would permit the development of assessments relating to specific factors or concerns within a more general analysis framework.

Modularization will force independent specific definition of assessments required for each major impact area. This should help clarify the judgments being made. While this does not solve the "value problem," it will clarify for government and public review the nature of the judgments made. The modular structure should facilitate "fine tuning" adjustments to assessment modules which necessarily result from the governmental and public review process. In addition, it should be possible to make iterative use of the system by looping back through module levels as appropriate.

Four distinct major phase modules can be identified and treated as system modules in the planning and assessment process for routing utilities: Data Base Selection and Development, Impact Assessment Modules, Evaluation and Composition of Assessments, and Routing Analysis. Data base selection and development includes not only the identification of appropriate information sources, but includes a subsystem to add information to a data base. This subsystem should be modularized to facilitate parallel task execution with systematic data certification and addition to a data bank. Development of assessment modules would simply consist of the independent (possibly overlapped) development of appropriate series of analysis routines which would be applied to the data base developed in the first phase. Evaluation and composition of assessments would consist of the calculation on a grid cell basis of each assessment module's cell rating and the composition of the ratings into a final assessment. This phase would be modularized into two system submodules: first the rating calculation and second the composition of the ratings. Routing analysis consists of the sequence of finding least-cost paths related to the composite impact assessment and analysis of the nature of resulting routes. This phase can be further submodularized.

In addition, mapping data and analysis would be an ongoing phase through most of a study. It would be itself modularized to provide the different maps required for data certification, output of analysis modules, and routing.

REQUIREMENTS OF SYSTEM PHASES AND CAPABILITIES

Data Base Development

Data base development is based on the map series available. It should be possible to maintain resolution of phenomena close to that available in the source document. Various map scales are in wide use; however, in planning applications only map scales of 1:25,000, 1:50,000, 1:125,000, and 1:250,000 are common. For large-scale regional planning purposes, scales of 1:125,000 and 1:250,000 are used. For detailed regional planning, 1:25,000 and 1:50,000 are the preferred scales; these scales would be used for moderate-sized projects. However, only selected urban areas are covered at the 1:25,000 scale. As a result, the most detailed map series which is available on a consistent basis is at the 1:50,000 scale. A survey of such sample maps reveals that its maximum resolution is 30 to 40 meters. The actual scale of phenomena mapped below 80 meters is imprecise. For example, woodlots below 160 meters square (*i.e.,* six acres) are not shown. These maps are often used as base maps for land classification schemes. An examination of many representative classification maps shows resolution of approximately 300 to 750 meters. This implies that a basic analysis grid size between 250 and 750 meters is sufficient in most applications providing some consideration can be made of detail down to 60 to 80 meters. This requires the development of a scheme of organizing data on a grid cell basis but maintaining details below that level. That is, the lowest level of addressable data is a single grid square; but, each grid square may be described by a series of percentage breakdowns.

While an arbitrary study will be standardized on some selected grid scale, certain data sources may only be available at other scales and some mapped output from the study may be required at different scales. Accordingly, development of the data base, analysis and mapped output must easily accommodate scale changes.

To provide reliable results, an all-inclusive and effective data certification procedure must be included. Some map series will consist of very detailed, others less detailed, coded data zones. It is essential that map codes can be used directly (without translation) and that the input process be capable of working at various levels of precision suitable for the nature of the map series. The system should accommodate data acquisition on a map series basis. Data acquisition, certification and correction should be possible in an overlapped manner and easily performed by nontechnical staff.

Mapping Facilities

Maps will be used for data certification and to demonstrate the results of study assessments, and alternate routings. Maps produced for data certification purposes should portray accurately at any suitable scale the existing contents of the assembled data base and should be produced quickly at low cost. They should be directly usable without major technical adjustment. Maps produced demonstrating results of analyses and routing exercises should be of high quality suitable for nearly direct reproduction in reports. Shading used on the maps and map scale must be flexible. It must be possible to add outlines of standard political units and the alternative corridors to the final assessment maps.

Impact Assessment Modules

It is essential that decision-making criteria which can be quantified easily be implemented when developing analysis modules. It should be possible to use, in combination, any of nominal, ordinal or interval scales. In addition, threshold calculations as proposed by Malisz (1972), and neighborhood functions as proposed by Peuker and Chrisman (undated) must be possible. Naylor *et al.* (1968) observe that complete flexibility is only achieved when the full facilities of a high-level general purpose programming language are available. Accordingly, analysis modules will be independently developed as subroutines and the data bank storage discipline will accommodate easy use by such subroutines.

Identification of the number and characteristics of assessment modules or factors should be dependent upon the nature of the utility under consideration. A general description of these factors should be made early in a study. The properties of the factors would be analyzed to determine the specific data items required to evaluate the factor. Analysis modules should not only include the traditional costs of construction, physical inventory, etc., but also recreation and cultural values, public and governmental concerns, and so forth. Each module should be developed separately from the others and should be independent in nature. This will help make clear the differences between the modules, and the lack of duplication or "double counting" will facilitate public and governmental review.

Module independence does not require, however, complete independence in data variables referenced. Analysis modules should use whatever data inputs are appropriate for the assessment required.

Evaluation and Consolidation of Assessments

Evaluation of assessments should be possible on an iterative process as assessment modules are refined. Certification of the correctness of these modules and subsequent study documentation purposes requires a facility to produce a high-quality map of all or part of the assessments. After assessment modules are functioning in a satisfactory way, it should be possible to add the results of the various assessment modules to the data base for reference in future studies.

The composition of assessments should be at the discretion of the investigators. While a good composition algorithm should be developed and available for general use, it should be possible to implement any arbitrary straightforward algorithm.

GENERAL REQUIREMENTS

In addition to the requirements outlined above, some more general ones involving data to be used, the value-ranking problem and the No Development problem may be outlined.

Data to be added to a data bank should be complete and consistent across the study area under consideration. It is desirable to include not only data derived from map series which are essentially professional interpretations or judgments but also direct data on physical features (*e.g.,* elevation, presence of standing water, etc.) and data obtained reflecting public or governmental opinions and concerns. This last data source could help in defining a "public concerns" analysis module. The presence of factual physical data can be used to certify accuracy of some interpretive data sources and to refine such assessment codes of various map series to establish a new interpretation. For example, an interpretive map series is available which assesses the capability of land to support quality woodlot stands whether or not woodlots exist. This interpretation could be intersected with physical data indicating where woodlots exist. The result would be an assessment of the potential quality of existing woodlots.

In order to clarify the judgments made and the data resources utilized, it is important that the system facilitate the production of partial results in a readily interpretable form. Maps of basic data, data composites, and results of individual assessment modules should be easily produced. In addition, a verbal description of the computational rules used to derive the maps should be made available. While this would not solve the value ranking problem, it would identify the judgments made for possible revision during the governmental and public review process.

The *No Development* or *No Go* alternative should be included where appropriate in module rules. It is anticipated that the study team would specify combinations of conditions which would give rise to such a high level of impact in a grid cell that the area should be considered *No Go*. An example of this could be the presence of a significant nesting area for an endangered bird species. In general, impact would be rated from No Impact to *No Go* or Prohibitive Impact. These assessments being completed at the individual factor assessment level would be combined to give a total assessment which would include no development areas. The actual attempt at routing would take place only after the impact assessment. If no paths can be developed due to *No Go* areas, this should designate a No Development project. The provision of summary analysis of proposed routes indicating the amount of various impact level land included will provide a basis for governmental review of project suitability.

CHAPTER 4

PLANNING SYSTEM AND ITS COMPONENTS

The nature of the new planning system and its components is now outlined. The system itself covers all phases from data collecting through assessment of impacts to identification and assessment of possible routes. An application of the system follows an iterative process as study parameters are refined. However, the fundamental sequence of events may be outlined as follows:

1. The nature of the utility and the study region are used to establish a set of impact factors to be assessed.
2. Data variables are selected to facilitate assessing the impact factors. A data base is developed for the study area on a small grid area basis. A new computer procedure is used to input data from map sheets.
3. A series of algorithms is defined for each of the impact factors utilizing data base information to determine impact assessments.
4. Impact factor maps are drawn by a special computer program driving a digital plotter.
5. Impact factor values for the independent impact factors are combined to provide a composite impact assessment. While this may be achieved by a number of arbitrary methods, a new cascade algorithm is provided to combine the (usually ordinal) impact assessments.
6. Composite impact assessments are mapped in the same way as the individual impact factors.
7. Corridor identification is based on the graph representation of composite impact. The graph is contracted and alternate routes selected by a modified Dijkstra algorithm. The alternate routes are straightened and corridor impact tabulated.

The total system is described in this chapter. The major phases and software algorithms are discussed in Chapters 5 through 10, and an application discussed in Chapter 11.

INTRODUCTION TO THE SYSTEM

A new general system was developed to support the full sequence of events required to perform an environmental impact routing assessment. The system itself consists of a procedure implemented by a professional analysis team wherein a number of computer procedures are applied on an integrated basis. This system finds application in one or more roles in the total policy development process. While the system itself is implemented by professionals it may be used by any or all of project proponents, government ministries, or intervenors. It is anticipated, however, that the inherent flexibility in the system would encourage one application of the system per project. This could be accomplished if the study implementation team takes care to include all of the significant parameters relating to each interest group. Regular interaction with proponent, government and specific interest groups, incorporation of data and concerns of each group, and public review will be required to realize the full value of this planning system.

The system is based in part on work done in a number of different areas. However, some of the fundamental impetus can be associated with work by McHarg (1971), Alexander and Manheim (1962), Rattray *et al.* (1973), Isard and Van Zele (1975), Krauskopf and Bunde (1972), Leopold *et al.* (1971), Turner (1970,1972), Fisher and Davies (1973), and Malisz (1972). McHarg has shown the advantage of intensity analysis on a precise spatial basis when evaluating placement of utilities. His use of color cartographic overlays identified the need to assess particular functional concerns or factors. Alexander and Manheim, Krauskopf and Bunde and Rattray *et al.* have established that McHarg's approach can be effectively replicated on a computer system by using a grid analysis technique. This is the fundamental assessment strategy used by the planning system.

The proposal by Rattray *et al.* to include "red flags" signifying a very special condition in a specific grid location has been utilized in various areas of the new system. Leopold *et al.* and Turner identify the need to consider intensity along with the basic assessment of coverage. In addition, their work emphasized the need for orienting the system to permit analysis by concerns or "factors." Newkirk and Dooley (1976) identify the value of these factors being orthogonal in nature. Both Turner and Krauskopf and Bunde have attempted route determination methods which served as useful reference points in developing the routing algorithms used in this system.

Fisher and Davies note the requirement to identify heavy impact conditions which arise from a series of moderate but cumulative impacts. They propose, in addition, as does Malisz, that impact thresholds be included in a study. The planning system was structured to permit this kind of analysis and has given rise to a new Cascade Algorithm.

PLANNING SYSTEM RELATION TO THE PLANNING PROCESS

The application of the planning system will usually be in reaction to a proposal by a proponent to establish a utility. The actual environmental routing analysis is only a part of a complex policy development process. To identify the relations between the planning system and the policy development process, a sample policy development scheme is shown in Figure 4.1. The planning system can find application in two areas: "Environmental Macrorouting Analysis" and "Detailed Routing Analysis."

The figure shows the "reactive" nature of environmental assessment to the decision process involving actual and estimated requirements, supply conditions and development strategies which have resulted in a project proposal. Many projects have involved an initial benefit/cost study and left the environmental routing analysis until after a policy decision to proceed has been made. It is proposed that an initial macroscale environmental routing analysis be performed to provide a realistic assessment of "costs" for input to the benefit/cost study. The planning system accommodates public involvement; this would facilitate effective public input in the policy development stage. A more detailed assessment would follow after the basic policy had been established concerning utility development. As shown in the figure, iterations may be required at the policy level.

Macrorouting analysis differs from detailed routing analysis in the scale of the unit area or grid unit used in the study. In addition, possibly fewer and less complicated impact factors are evaluated. This kind of study could be applied simultaneously to a number of possible study area configurations to help establish the potentially least-cost area for utility development. For example, such a series would be performed before settling upon an area like Canada's Mackenzie Valley for utility development. A major advantage of the macrorouting analysis is the provision of alternate approximations of least-cost routings.

INTEGRATION AND CONTROL OF THE PLANNING SYSTEM

The planning system consists of four distinct analysis phases in which various inputs of information, data sources and computer procedures are coordinated by a professional assessment team. Figure 4.2 shows the

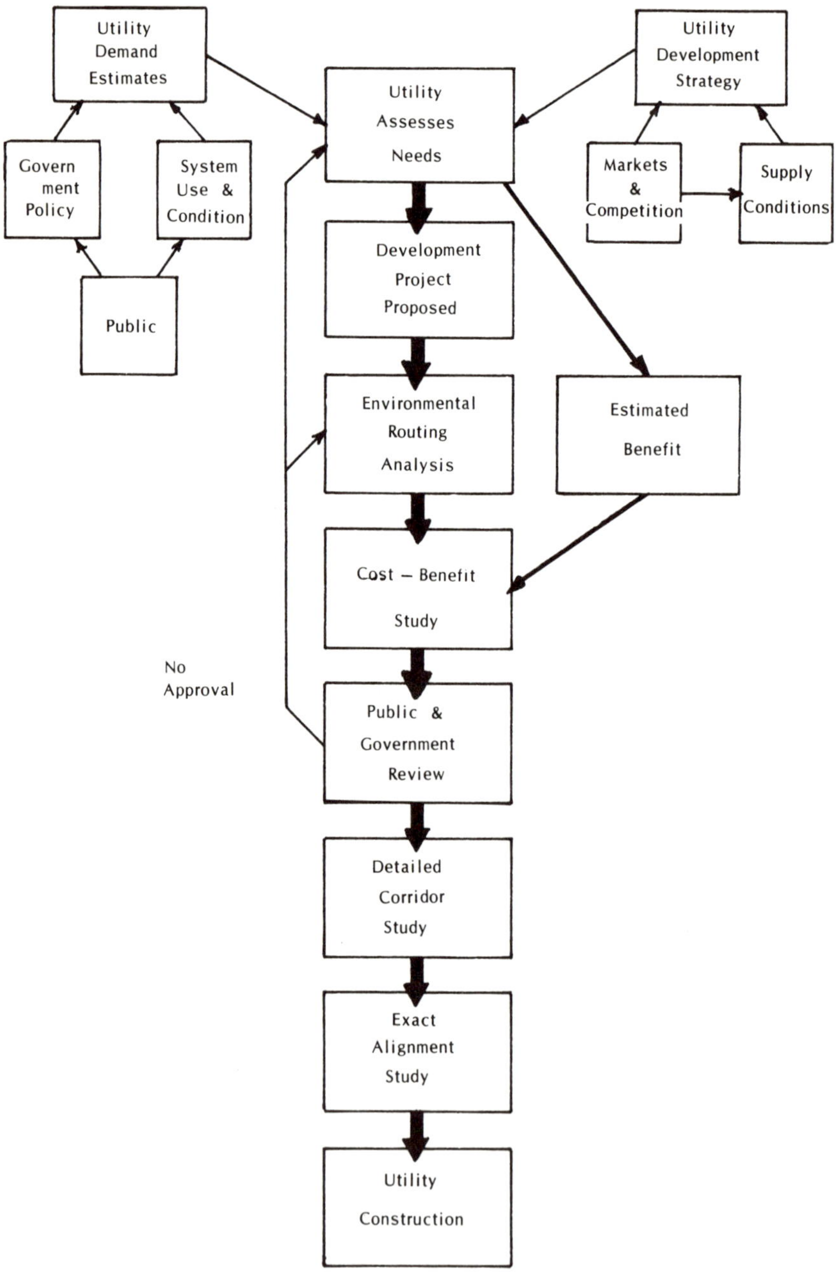

Figure 4.1 Policy development.

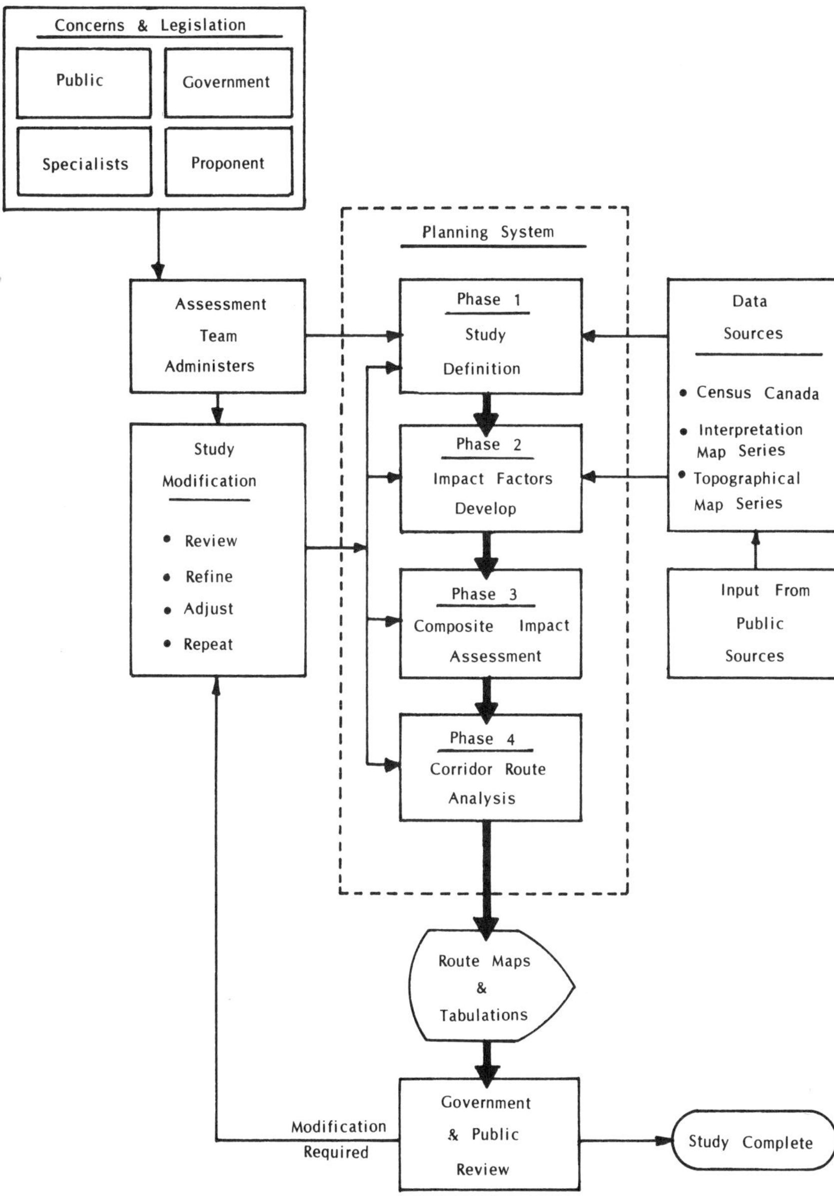

Figure 4.2 Planning process and system integration.

general relationships in this process. The assessment team could be assembled from proponent's staff, from a governmental agency, an organized intervenor group or a specialized assessment agency. Quite possibly, the team could be assembled by selecting experts from each group. In practice, it is anticipated that the team will be assembled and coordinated by a consulting firm in the private sector retained either by the proponent or by the government. The assessment team serves as the interface between the analysis system and all interested parties. Needless to say, the funding support received by a study and the quality and performance of the team will directly affect study adequacy. To facilitate full interaction with all interest groups, it is essential that the assessment team function relatively free from direct influence of any interested party. The public, government and proponent reviews indicated in the figure are designed to ensure that the assessment process includes their essential concerns.

Initial definition of the planning analysis is the responsibility of the assessment team based upon legislation, guidelines, specialist advice, specific directives of governmental agencies and information from the proponent and the public. This last source of defining information could be obtained from public surveys of various kinds. Based on this initial definition, partial results of an assessment are produced for full review by the interested parties. Requests for revisions and adjustments which result from the review are used by the assessment team to adjust analysis system operations. The analysis system is subsequently reactivated at the appropriate part of the earliest phase required, and a new set of partial results is produced for review. An *iteration* of the assessment consists of going from one set of partial results to the next. The study proceeds through a series of iterations until it is judged acceptable. Usually this judgment is the prerogative of the appropriate governmental review board or ministry. In many cases, study and project approval may require a political decision by the government.

The time required for each iteration will depend upon the complexity of the revisions required and the nature and availability of data to be added. Generally, the requirement to repeat at an early phase is more expensive in time and cost than to repeat at a later phase. At any iteration, additional data and impact assessment models can be added in addition to adjustments to existing analysis parameters. This kind of feedback iteration is common to the general planning process and is similar to control problems in chemical plants as shown by Light and Johnson (1974). Such a control system must iterate until the appropriate parameters are determined for each input stream. The number of iterations is directly related to the initial parameter setting and the appropriateness of each subsequent correction. Accordingly, the initial study parameterization by the assessment team (*i.e.*, phase I) is crucial.

OUTLINE OF PLANNING SYSTEM PHASES

There are four major phases in an application of the planning system. They serve as major analysis models. Figure 4.3 shows the basic components of the phases and their interrelationships. Within a specific system phase or phase component there is often a procedure or a series of submodules. Detailed description of system phases and their components will be found in subsequent chapters.

Phase I: Analysis Definition and Structuring

The main features of this phase are: the selection of the major impact factors to be considered, the definition of algorithms to evaluate the impact associated with each factor and the development of an appropriate data base on which to apply the algorithms. The nature and characteristics of the proposed project, and relevant concerns, knowledge, and attitudes of the public and government are analyzed by the assessment team to determine a small set (*e.g.*, six to twenty) of major impact concerns or factors. These factors are examined to identify the important data variables which are required to assess each factor. This list of variables is developed in the context of available data sources to determine extra data tabulations or map series which must be obtained. A common *unit area* or grid cell size for data bank development and impact factor evaluation is selected. Additional map series are compiled and added to the data bank using data interpolation programs. Quantitative rules to be used in impact factor evaluation are developed in maching processable format.

Phase II: General Impact Evaluation

This system phase is entered only after phase I is complete since the data bank and the impact assessment rules developed in the first phase are required. This second major system module involves the detailed impact evaluation of each impact factor by individual unit areas. Following this impact analysis over the entire study area, a computer-drawn map is produced for each impact factor evaluated. A subject area specialist on the assessment team, possibly in consultation with an outside expert, reviews each impact *factor map*. The specialist backs up his review with actual field studies to assess the correctness and suitability of the impact assessment. The impact factor algorithm is adjusted, reapplied, remapped and reevaluated in an iterative process until a satisfactory assessment is achieved. In an extremely difficult case it could be necessary to revert to phase I and add more data to the system or perhaps restructure the major impact factors. It is anticipated, however, that adjustment to the

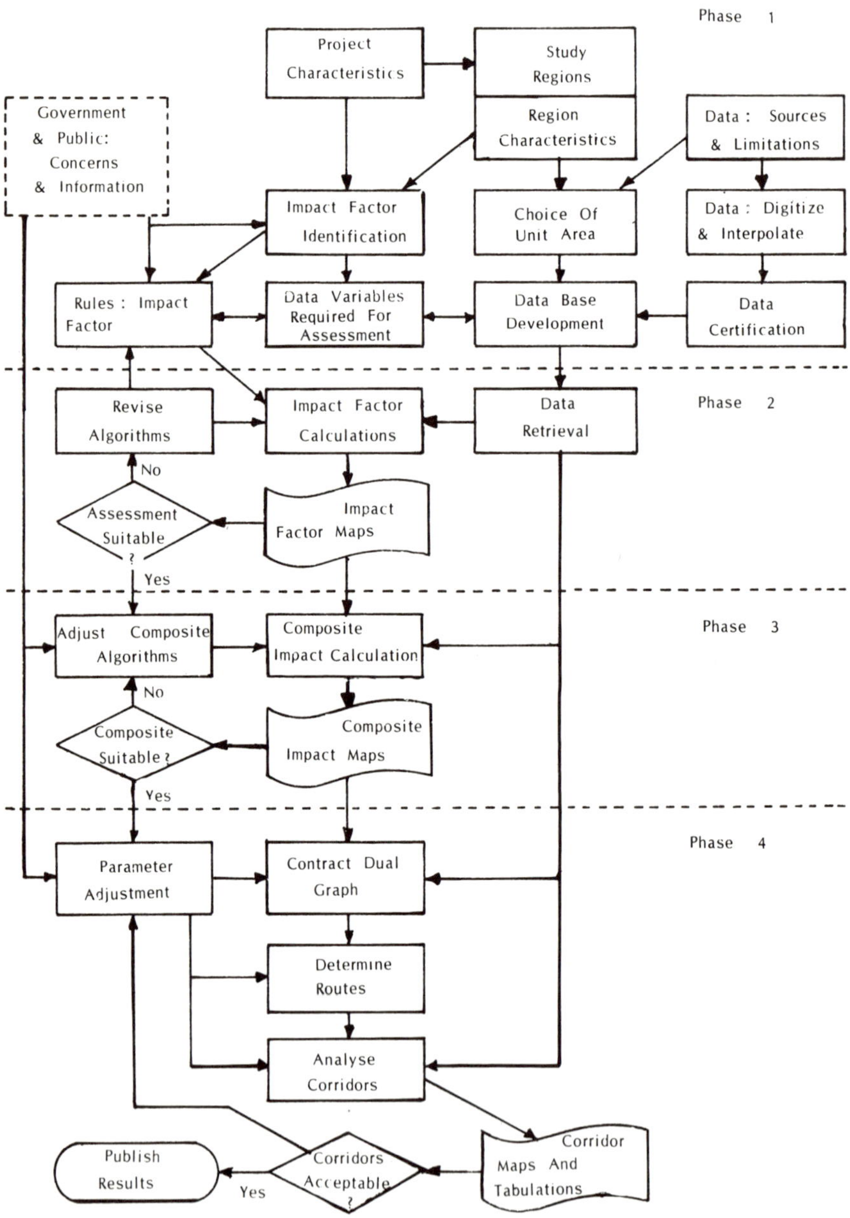

Figure 4.3 Planning system.

parameters in the factor algorithm will be sufficient to obtain the desired total assessment in most cases. Once all impact factor evaluations are accepted by the team, the evaluation results are added on an individual unit area basis to the data bank as evaluation variables. The final maps produced in this phase are made available for external review by the public and government. Mapping precision is on a unit area basis and permits the precise identification of regions of high or low value or impact for the associated impact factor.

Phase III: Composite Impact Evaluation

This phase begins only after phase II is completed. Its purpose is to provide a single synthesis or "composite" of the individual impact factor evaluations. This crucial phase is required to establish on an individual unit area basis, and considering all impact factors, the severity of impact which would be associated with development of the utility. The assessment team is free to apply whatever scheme appears useful for combining the impact evaluations of phase II. Essentially the methods will be one or a combination of weighted averaging, threshold scoring or application of the Cascade Algorithm especially developed for this planning system. Any combination algorithm has a set of parameters associated with it which must be pre-set by the assessment team.

The composite algorithm is applied using the impact evaluation variables in the data bank; a single composite assessment variable is developed and added to the data bank on a unit area basis. A map similar to the impact factor maps of phase II is produced showing the results of the composite assessment. This map is reviewed for its correctness and suitability by the full assessment team (possibly expanded by the outside experts). In the event that the composite evaluation is unsatisfactory the team readjusts the composite algorithm or its factors, reapplies the algorithm, remaps, and reassesses suitability in an iterative process until the results are accepted. In an extremely difficult case, the team can revert back to phase II and a reassessment of the set of (or individual) impact factor evaluations. This could give rise to a phase I reevaluation of data resources and impact factor algorithms with the requirement to add more data and impact factors. It is anticipated, however, that most phase III revisions will involve only parameter changes in the composite algorithm. The final composite map is made available for public and governmental review.

A series of composite and subcomposite maps could be developed. They could be obtained by using different composition algorithms or impact factor combinations. Each could reflect a group's value judgments. The value judgments would, however, be concretely defined by the associated composition algorithm and its parameters.

Phase IV: Route Development and Analysis

This study phase begins only after phase III is complete since it is based on the composite impact assessment. Its purpose is to locate a set of alternate route corridors of arbitrary width which join the two points between which the utility is to be placed. A procedure is followed which attempts to find corridors with least environmental cost based on the assessment performed in phase III. It is, of course, possible that no corridor exists which has low environmental cost. In addition to locating a set of corridors, a detailed summary is provided indicating by area the impacts associated with each corridor.

The analysis procedure requires route development parameters to be pre-set by the assessment team. These include source and destination of the utility, corridor or right-of-way width, and a series of penalty parameters to be associated with impacts during both the route determination stage and the impact analysis stage.

A series of computer programs use the pre-set parameters and the data base to develop corridors. The corridors are drawn as an overlay by the computer mapping program and tabulations of impacts associated with each corridor are displayed for review. The assessment team reviews the corridors for correctness and suitability. If required, parameters can be reset to produce a modified corridor configuration. This is particularly useful if either source or destination could be sited at various locations. Once the corridors have been finalized, they are drawn as an overlay on the individual impact factor and composite maps. These along with the associated impact tabulations are presented for public review.

REVIEW AND REVISION OF ASSESSMENT

A complete iteration of the planning system provides a set of impact factor maps, a composite map, a set of alternate corridors and tabulations based on a set of algorithms applied to a specific data base. The maps and tabulations are published along with a description of the algorithms and the parameter settings applied. As indicated in Figure 4.2, the public, government and proponent review and comment upon the total assessment. The team reviews any substantive criticisms of the assessment and may revert to any one of the planning system phases for revision.

As indicated in Figure 4.1, a government assessment follows the application of the planning system and may cause the study to be repeated under special instructions or approve it for further implementation.

CHAPTER 5

PHASE I: ANALYSIS DEFINITION AND STRUCTURING

Phase I establishes the foundation for an application of the planning system. The study data base and the impact assessment rules which support the remainder of an analysis are developed. The nature of the data base is determined by the selection and definition of a set of broadly based impact concerns or factors. The general sequence of events in this phase is shown in Figure 5.1.

SELECTION OF IMPACT FACTORS

An impact factor is used to represent a major sector of the environment that would be affected by the proposed project. They are essentially mappings from a specified subset of the study data base to an arbitrary set of numerical impact ratings. The professional assessment team analyzes the nature of the proposed utility, the relevant legislation and information from specialists and nonspecialists alike to determine the set of impact factors to be considered in a study. Factors are selected to provide a minimal nonoverlapping and exhaustive set of factors relevant to the study region and the project under consideration. For example, in a recent study on locating a high-voltage electricity transmission line (Newkirk and Troughton 1975), the following factors were considered: agricultural operations; natural environment; recreation, cultural and historical; residential, institutional and commercial; established linear utilities; area utilities; natural diversity; and relative visibility. Newkirk and Dooley (1976) report the value of establishing factors which use subsets of input data variables which are as disjoint as possible. This reduces the risk of double counting and makes clearer the differences between factors.

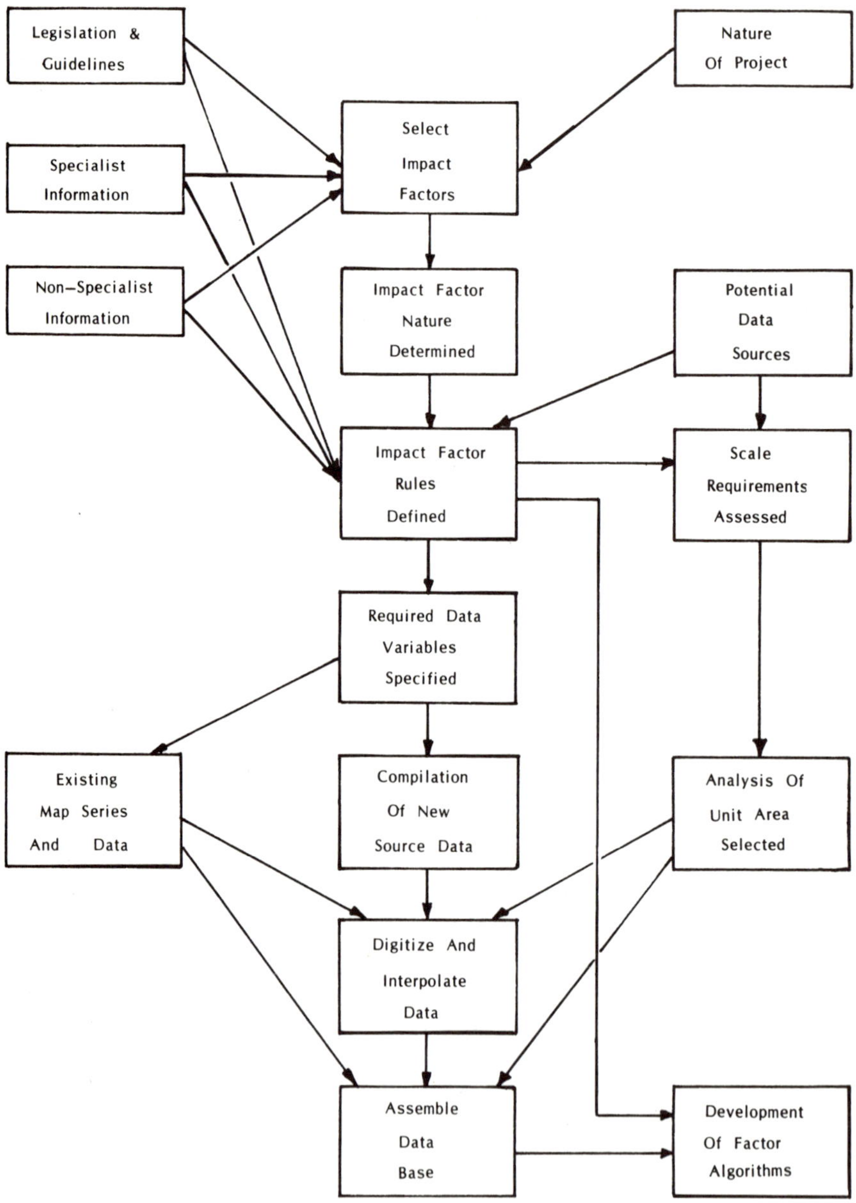

Figure 5.1 Sequence of phase I events.

Factor Identification

An impact factor is usually identified with a single major sector of the environment or man's activities or with a collection of sectors which are complementary in nature. This permits internal consistency in the nature and use of the data variables required to perform the impact assessment for the factor. In selecting a set of factors, future discussion at public and governmental review is facilitated if factors are selected which address directly sectors which are likely to involve controversy. This helps to avoid having the discussion of issues clouded with "irrelevant" data and criteria. Various methods or combinations of methods are used in system development to help identify impact factors.

Checklists of major assessment areas to be considered have been developed for various utility projects under the auspices of the Environmental Protection Agency and various government agencies. These can serve as useful guides in establishing appropriate clustering of concerns to establish the basis for an impact factor. However, care must be taken to identify impacts which may be local in nature and therefore not included in the general checklist.

Interaction and dependency matrices are very helpful in identifying significant impact sectors. A matrix is set up with columns assigned to each major action to be associated with the project. Rows are established (based on general checklists) listing all possible segments of the environment which could be affected by any action. Where an action is identified as having environmental effect an entry is made in the table with its magnitude corresponding to its intensity. Numerical signs are used to indicate direction of effect. Good examples of this kind of table are found in Leopold *et al.* (1971), and Lyle and Von Wodtke (1974). These matrices can be analyzed later to identify clusters of impacts with similar characteristics.

Diagrams and directed graphs may be used to describe the relationships between actions associated with a project and specific environmental effects. In addition, interrelations in environmental effects may be included. The resulting graph is then examined to determine sets of vertices with similar characteristics. Graph theory techniques of locating "strong components" can be applied to locate sets of vertices (*i.e.*, actions and effects) which are strongly interrelated. These strong components can serve as a first approximation of impact factors.

Surveys can be useful in selecting a set of impact factors. The public, local governments and interest groups can be administered with special surveys eliciting rankings of various environmental concerns, specification of interrelationships and so forth. The surveys can be organized to establish entries for interaction and dependency matrices. Cluster analysis can be used to detect clustering of "similar" items. A common clustering model minimizes within group variance on a set of variables and maximizes between group variance. Impact factors which are developed from this basis will tend to be disjoint in nature one from another and internally consistent. Where questionnaire results contain mainly interval values which have near normal distributions, the techniques of factor analysis may be applied to develop estimates of impact factor models. An added benefit of this technique is the estimation of relative importance between the "factors."

Once impact factors have been identified, their major components must be specified along with the data items required to evaluate them. The relationships between these concepts may be seen in Figure 5.2. A narrative description of the broad nature of each factor is prepared by the assessment team. Each factor narrative is given to subject area specialists who develop an outline of the factor by specifying its significant features and the data items required to assess the features. This is based in part on the results of the surveys which were conducted as well as specialist technical knowledge. Impacts with temporal dimensions are included as well as any significant side effects. As much as possible, the organization of a factor is based on a tree structure similar to the example shown in Figure 5.3.

Once the tree structure for an impact factor has been developed, subject area specialists develop rules for establishing the value of each branch of the tree. This includes associating an appropriate weight for each branch of the tree. The task requires careful professional judgment based on published guidelines, legislation, information from specialists and input obtained from the public. The values and weights assigned are applied consistently across the study area. This forces the expert to make as objective an assessment as possible and to provide exact specification of the values and weights to be applied. Part of this exercise involves selection of the method of combining the weighted results to derive the final assessment for the impact factor. A list of variables is developed which the specialists place in order of importance for impact factor assessment. The list is compared against the available data resources. Where possible, the factor assessment model is adjusted to maximize its use of available data and minimize the requirement for new data while maintaining accuracy. The outcome of this process is a priority ordered list of data variables ranked essential, useful, and peripheral to impact factor assessment.

Let $(y_1, \ldots, y_k)$ be a set of measurements for "k" impact concerns or factors $Y_1, \ldots Y_k$. In general $y_i \in R$ where R is the set of real

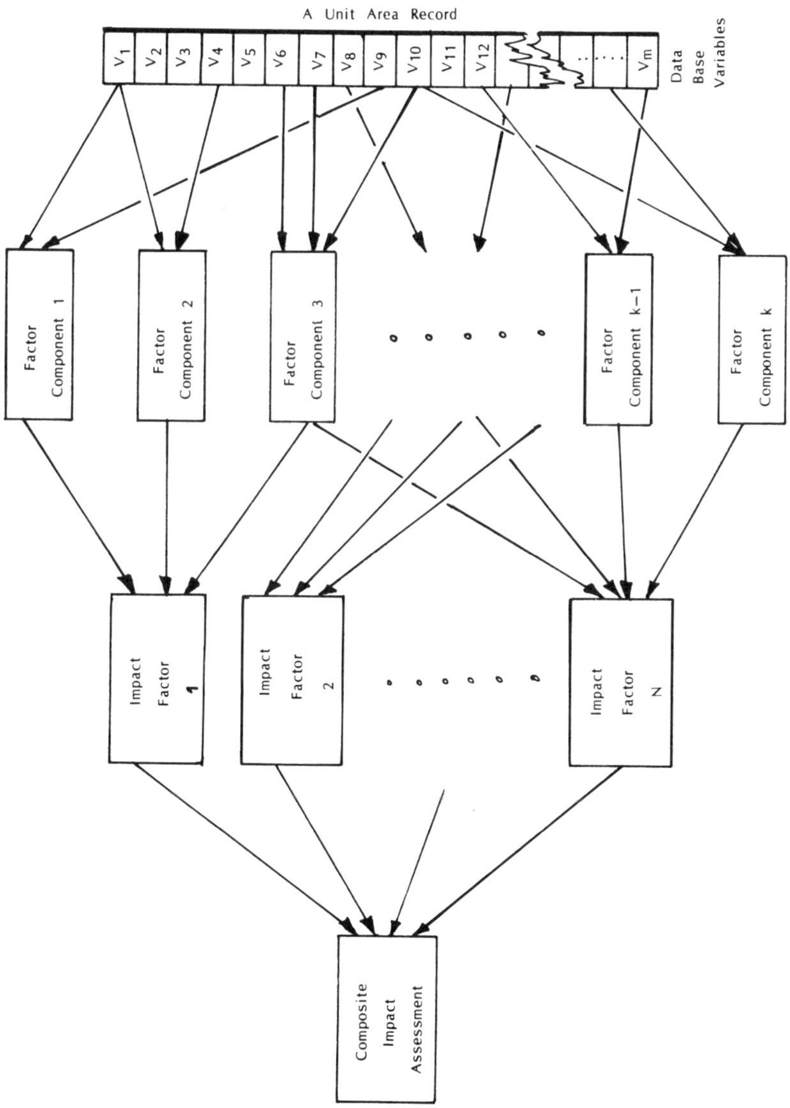

Figure 5.2 Composite impact, factors, components, and data base relations per unit area.

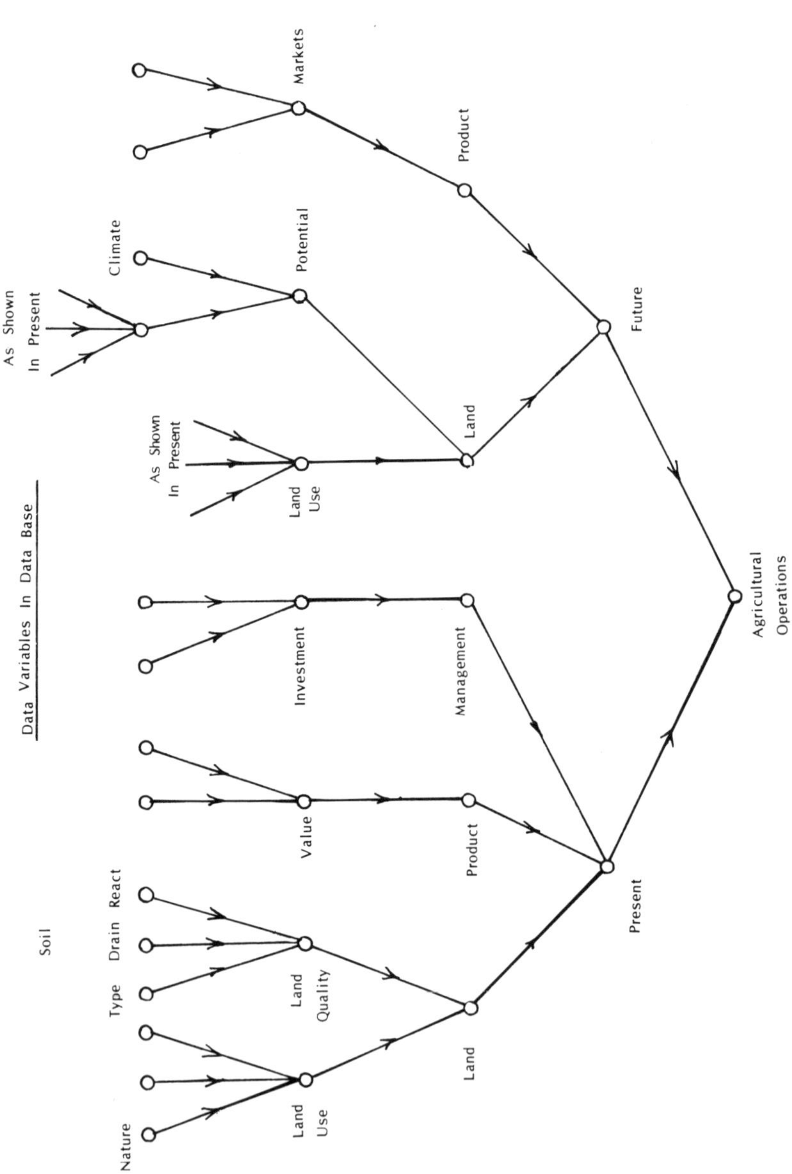

Figure 5.3 A sample impact factor structure.

numbers but typically y_i will be an element of a subset of R such as the nonnegative integers or some bounded set of positive integers.

The choice of measurement units for y_i will depend on the nature of planning, concerns or factors and the opportunity that is afforded to use commonly accepted measurements for those concerns. An obvious example is to use monetary units to measure economic concerns. For many factors there will not exist a natural measure; it is common in such cases to use an arbitrary scale of dimensionless units or rankings.

It is convenient for composite impact assessment that all impacts are measured in terms of an interval scale of integers $Z = \{1,2, \ldots ,\} r$ when "r" is some arbitrary maximum score for Z. This may require a transformation of the natural measure of impact. In making this transformation, any differences in relative importance of factors can be accommodated by scaling the natural measurements to appropriate intervals $1,q$, $q \leqslant r$.

A distinct characteristic of the study area will be denoted by X_i and referred to as a variable. Present land use is an example.

For $i = 1, \ldots , k$, let $B_i = X_{il}, \ldots , X_{ip}$ be a set of variables having values

$$x_{ij} = x_{ij} | x_{ij} \text{ is a value of } X_{ij}$$

so that

$$(X_{il}, \ldots , X_{ip}) \in X_{il} \times \ldots \times X_{ip}$$

Denote the latter produce space by W_i.

A factor algorithm is a mapping

$$U_i : W_i \rightarrow Z$$

such that the inverse image of Z partitions W_i. There are some further important properties of U_i that are not readily expressible in mathematical terms. Specifically there should be a blend of scientific basis, judgment and ecological acceptance of the mapping U_i. The important point is that the credibility of the information presented to the decision-maker depends on the validity of the mappings U_i, $i = 1, \ldots , k$.

The data base composition is

$$D = \bigcup_{i=1}^{k} B_i$$

and the data base is the set of values for D for each unit area of the study region. It is not necessary that

$$B_i \underset{i \neq j}{\cap} B_j = \phi$$

although the closer the intersection order is to zero the easier it is to interpret the Y_i's relative to each other. This is so because if the intersection is not empty, then at least one variable is used in more than one factor rule which could be interpreted as double counting. In general, repeated use of a variable should deal with different aspects of the characteristic.

The mapping U_i can be quite general. For example, consider, the factor "impact of project on residential commercial and institutional land use." This might use the characteristics "present land use," X_1 and "planned land use," X_2 where $X_1 = \{$urban residential, rural residential, commercial, institutional, industrial, agricultural, recreational$\}$etc. and X_2 has a similar set of values except that they have values in the data base for planned land use instead of present land use.

Suppose that a mapping V is defined on X_j for j = 1,2 as follows:

X_j	$V(X_j)$
urban residential	5
rural residential	4
commercial	3
institutional	2
other values	1

In this case $V(X_j)$ expresses the impact of the project on various forms of land use related to residential, commercial and institutional use.

If the trade-off between present and planned land use is to be in the ratio 7:3 then the factor rule would be

$$U\ (X_1,X_2) = 0.7\ V(X_1) + 0.3\ V(X_2)$$

This simple example also illustrates that other aspects of a variable could be considered without double counting. Specifically, the impact on the factor "impact on agricultural operations" could make use of the values "agricultural" in X_1 and X_2.

INITIAL DATA BASE SPECIFICATION

This step includes the selection of variables to be placed in the data base, their sources and the organization of data related to scale and accuracy. The team assembles the priority ordered lists of variables from each impact factor, and develops a combined priority list. Priorities are assigned based upon both the number of factors requiring the variable and the variable's relative importance to each. The final list of variables is reviewed by the specialists responsible for each impact factor. The specialist indicates for each variable the spatial precision required. When data are available in a hierarchy of class codes, the level of class code required is also specified. The assessment team augments the variable list by indicating for each variable the greatest amount of precision required. The augmented variable list is scanned by the assessment team to determine the unit area size required to provide adequate localization of phenomena.

Data Base Organization and Size

The unit area (*i.e.,* grid unit) provides the basic analysis unit for information organization in the planning system. The unit area represents a square area in space. Unit areas are nonoverlapping and exhaustive of the total study area. Each is the same size and is uniquely identified by the geographic coordinates which locate the point at its center. A separate data record is established for each unit area in the study region. The size of the unit area limits the maximum resolution (or localization) of study region features and impacts. The use of square regions gives the same degree of localization in both dimensions. For example, the larger the unit area, the less accurate an assessment or environmental phenomenon will be when subjected to a spot-specific field check. The unit area need not be considered as homogeneous with respect to any phenomenon although it may be convenient to do so. When it is considered as homogeneous, the unit area is usually assigned the characteristic of the phenomenon which explains the largest percentage of the unit area. The characteristic recorded for the unit area can be said to be located at $(X \pm r, Y \pm r)$ when X and Y are the unit's geographic coordinates and r is one-half the length of one side of the unit area.

A unit area not considered homogeneous for a specific phenomenon must have recorded the various characteristics observed for that phenomenon. Following the work by Krauskopf and Bunde (1972) this is achieved by recording the appropriate characteristic code and its corresponding percentage of the unit area. In many cases it is only necessary to record some arbitrary number of characteristics which explain a maximum percentage of the unit area. These can be entered into the file in order of percentage area explained. This facilitates quick location of the most significant, second most significant etc. characteristics detected. The degree of spatial localization remains the same as for the homogeneous phenomenon.

When a maximum number of characteristics to record is specified, it may be necessary to make a record of an arbitrarily defined characteristic no matter how small a portion of the unit area it occupies. These characteristics can be called "red flag" data items. For example, it may be important to record the presence of a historic site in a unit area even if it has only a very small spatial extent. Such red flag entries can be added to the data record for a unit in an arbitrarily assigned position.

Unit area size is selected to achieve the maximum degree of localization required analysis for the study (*i.e.*, to reduce "r" to the maximum acceptable value). The number of basic analysis data records in the study is dependent upon the size of the study area and the length of the side of a unit area. As unit area size is decreased, the number of records in the data file increase as the square of the ratio that the new unit area side length is to the old. Accordingly, a study must be scaled to a value of "r" which provides required precision and a data base which is not too costly to store or analyze.

Data Base Structure

Geographic coordinate values serve as the keys to data records. They are recorded as the first two variables in a unit area record. In an application of the planning system, the evaluation of impact factors, development of composite impact assessment, route analysis and mapping requires individual analysis of every unit area record. Accordingly, it is sufficient to process the data base as a sequential file. To facilitate merging various data files into the master data base, the sequential data files are ordered by coordinate. The file is ordered from north to south as the primary key and east to west as the secondary key. If the data base was to be used later for inquiry purposes outside the planning system, it would be convenient to "invert" the file on the geographic coordinates using an index sequential access method of some kind.

The number of data variables recorded for a unit area will vary from study to study and may vary during the life of a study. In addition some unit areas may require more and some fewer than average number of variables to record its nature. The data structuring required for units with varying information content is shown in Figure 5.4. Record "p" is shown located by its geographic coordinates. It has a set of fixed record items and some variable content items which are linked to the record. The next record is linked via its geographic coordinates. It must contain the same number of fixed record items as the other records. However, as shown it has fewer variable fields. The first variable part has only two entries while for record "p" there were more than four. The second variable part of the record is shown to be null.

The list structure shown in Figure 5.4 can be effectively implemented using variable field, variable length records. A sample record layout is

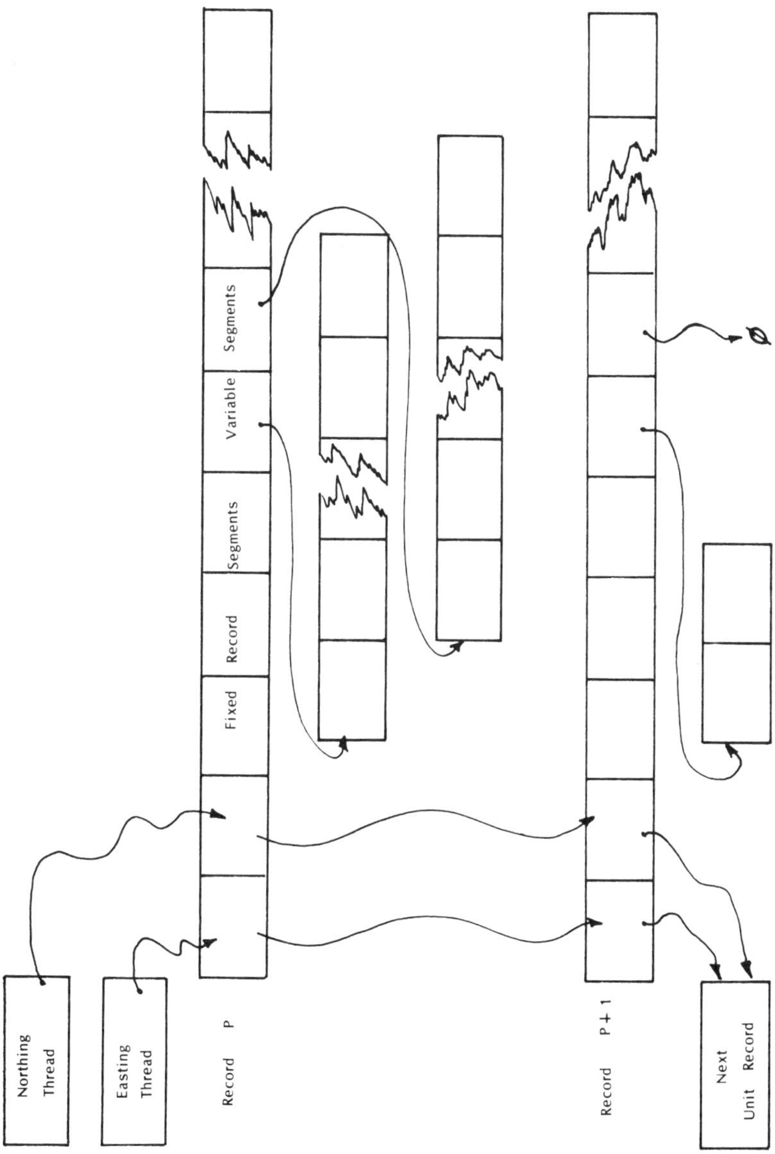

Figure 5.4 Structure of a unit area record as a list.

shown in Figure 5.5. Record positions 1 and 2 hold the geographic coordinates which serve as record key. Positions 1 through 9 hold fixed position information which is always coded for a unit area. This is also true of positions p+1 through p+10; the positions of this latter set are always fixed relative to the last land use variable field. Note that it is possible to use arbitrarily interval, ordinal or nominal data as appropriate. For example, record positions 1,2,3,10,12,14,p,p+2 and p+4 contain interval values, positions 4,7 and 8 contain ordinal values and the remainder contain nominal values. The land use subrecord has variable length; the value in record position 10 indicates its length. Positions p+5 and p+6 have special coded values assigned if certain cultural or spawning-nesting areas are detected even if they have very small spatial extent. Positions p+7 through p+10 are assigned values only after phase II and phase III of the study are complete.

DATA BASE DEVELOPMENT

Once the basic variables required to evaluate the impact factors are defined, the study data base is assembled from existing and new data sources.

Data Sources

Data for input to the study data base may come from many sources. For example, existing computer data bases, cross-sectional or time series data for aggregate areas from a statistical agency, and existing map series can provide the source data base. Major difficulties involve determining the quality and consistency of the sources, and altering scale to that of the proposed study. In many cases published data tabulations and maps represent a higher level and more generalized or abstracted source and detailed source manuscript data or maps are not available.

Altering scale of existing map data in an upward scale is not too difficult since aggregations can be easily achieved. However, it is usually very difficult to do the reverse. To do so requires a statistical extrapolation process which takes into account local distributions. Tobler (1970) provides some computer programs which can assist in this process. However, they assume a continuous metric and cannot be satisfactorily applied to nominal and ordinal data.

In some cases it may be possible to avoid adding new data series if a variable can be calculated from already existing variables. This technique can be used effectively to improve the quality of an information series. For example, suppose it were necessary to have an identification of very

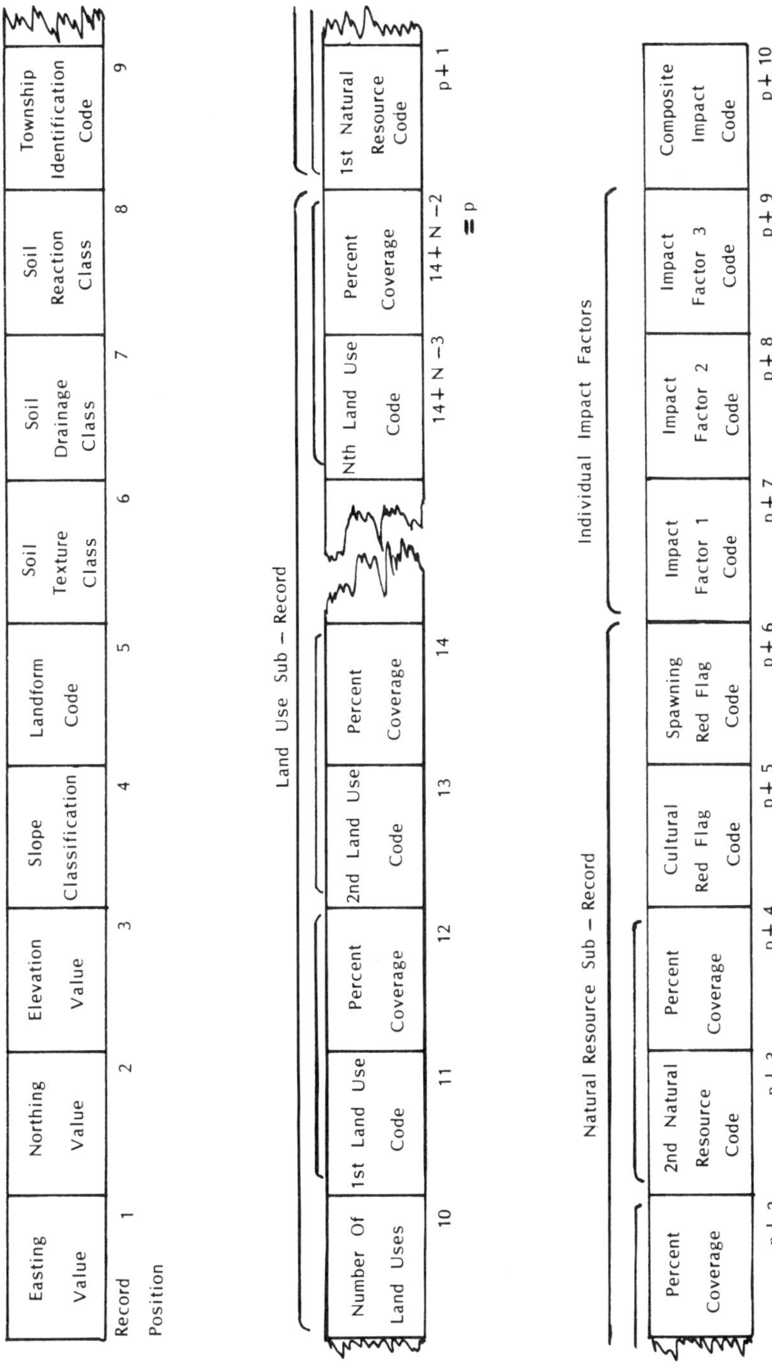

Figure 5.5 A unit area sample data record.

high-value woodlots in the data base. Suppose a general classification of all unit areas was available giving "capability to support woodlot," another was available indicating the presence or absence of actual woodlots, another was available indicating the proximity of first-order streams, another was available giving an assessment of stream quality, another an assessment for supporting upland wildlife, and so forth. The classification for each of these data series could be intersected to locate existing high-quality woodlots in head water areas of high-quality streams. Such a series intersection can be utilized during data base development. Data source intersection can often be effectively used to reduce the need for costly compilations of new data series.

Data Base Generation

A format is established for a unit area data record (similar in concept to the example in Figure 5.5). All variables to be added to the data base are assigned fixed positions relative to the start of the record or relative to some other fixed location. After accomplishing the appropriate aggregations or extrapolations to accommodate scale changes, existing data in machine readable form is added to the assigned variables in the file. Simultaneously, project staff compile new data tabulations and maps to provide the extra data required for the study. All mapped data and new tabulations must be put in machine-readable form for addition to the data base. This requires interpolating a value or series of values for each unit area record in the study area for each new map series. Given that a 2,700-square mile study area has roughly 40,000 unit areas, of 500 meters to a side, this can be a prodigious task. A method has been developed to permit a computer to perform the major portion of this map reading task; a detailed description will be found in Chapters 6 and 7.

In large studies, mapped input data cover a number of maps of a specific data series. Map sheet information is placed in machine-readable form by single sheets. A complete series of map sheets is input in sequence. Once input is complete it is necessary to combine the data records from all series and adjust the series "mosaic" to represent properly unit areas whose spatial extent spans two or more map sheets. The merged map sheet file is sorted by geographic coordinates; this brings the records for fragments of unit areas together. The fragmented records are then combined into one record using an appropriate linear or threshold model. Once a map series data set has had its mosaic adjusted, it is added to the data base in the appropriate variable locations.

Data Base Certification

Data base quality is the most crucial parameter in a planning system application. The best impact factor algorithms cannot overcome incomplete or incorrect data. While it is often easy to detect errors in an impact factor algorithm (because of its uniform application across the study area), it is not always easy to detect a data error. Accordingly, it is essential that all data in the data base be subjected to effective certification tests. A special certification step is included during computer map sheet interpolation. Five additional methods are applied to the full data base, the first of which is the detailed computer mapping of data base variables for review by a subject area specialist. This is effective at detecting both widespread and locally severe errors. The second technique is to apply a simple record completeness and redundancy check on the data base. This assures that all fields which should contain data items have values assigned, that a unit area record exists for each unit area in the study, and that there are no duplicates. The third technique is a base-bounds check of each variable in the study for all unit areas. Here the codes used for each variable are compared against the range or exact list of permitted codes to locate exceptions.

Following corrections of errors detected by the previous tests, the fourth test is applied. In this test, selected variables are compared against each other to check for inconsistencies. For example, if the soil texture variable indicates solid blue clay, the drainage variable indicates poor drainage and the agricultural capability classification indicates first quality farm land, a data inconsistency has been established. The fifth and final certification test applied to the data base is to perform a complete examination of each data field for randomly selected unit areas. The data values are checked against the relevant source documents. In addition to checking randomly selected unit areas, it is customary to inspect small groups of records in parts of the study area where data items are known to be complex.

When a record containing an error is detected, it is standard practice to examine the entire record for that unit area to correct all errors for that unit. All errors detected by a test are corrected before the next level test is applied. While the process may take a period of time, the importance of this stage in phase I cannot be minimized. It should be realized that the special checking which takes place in the computer map sheet interpolation (and is described in the following chapter) should minimize errors in the data base.

OTHER DATA BASE CONSIDERATIONS

Aggregate census data can be included in two basic ways. By using appropriate statistical disaggregating approximations values can be estimated for each unit area. If unit area detail is not essential, it is often useful to use the census data in an "indirect" way. In this case, a table is developed using the standard aggregation units used by the census agency. The actual values of selected census variables are stored in a table by census area. A map of census areas is input to the data bank so that each unit area has designated in its record the identification code of the census area to which it belongs. It is then possible to use the census area identification stored in the unit area record to perform a table look up to locate the appropriate census characteristic for that area.

Data base development and certification need not always be as complex and time-consuming as described in this chapter. As successive studies take place in a region, a regional data bank can be assembled. Indeed, it should be possible to construct a data base which is adequate for assessing the environmental routing problem for any major utility as long as sufficient data variables were maintained in the original data base.

CHAPTER 6

DATA BASE DEVELOPMENT

Data for a variable can be either point- or area-specific. Point-specific information is of the kind: "at point (X_i,Y_i) there is a microwave tower" or "there is a survey benchmark at (X_j,Y_j)." Area-specific information is of the kind: "the area bounded by Hickory Street, Lansdowne Avenue, Michaelmas Drive and Queensland Way is zoned for single family dwellings." The result of most land classification exercises is the designation of areas having specified characteristics. One can consider most urban and regional planning as a complex land classification exercise where region classes indicate suitability for a specific type of development or use. In planning applications, data variables with area distributions are the most common. An inventory of point-specific information is relatively easily maintained in the form of a series of lists of phenomena and their locations. Area data can be maintained by recording details of the boundary which contains the area or by listing the characteristics of the area as being associated with the set of subareas which compose it. Popek and Kingston (1972), Goodchild (undated, 1973a, 1973b), Peucker (undated, 1972), Steiner (1972), Steiner and Stanhope (1972) and Stinton (1972) remark on the difficulty of organizing area data for efficient analysis. They agree that planning applications require data storage and analysis on a uniform grid square or unit area basis similar to that used in the author's planning system.

Figure 6.1 shows an example of a land classification for an arbitrary area. Figure 6.2 shows the area when the unit area grid is laid down over the classification map. The problem is to establish what individual unit area codings should be to accurately represent the land uses coded on the source document. Source documents can have meaning as shown in Table 6.1.

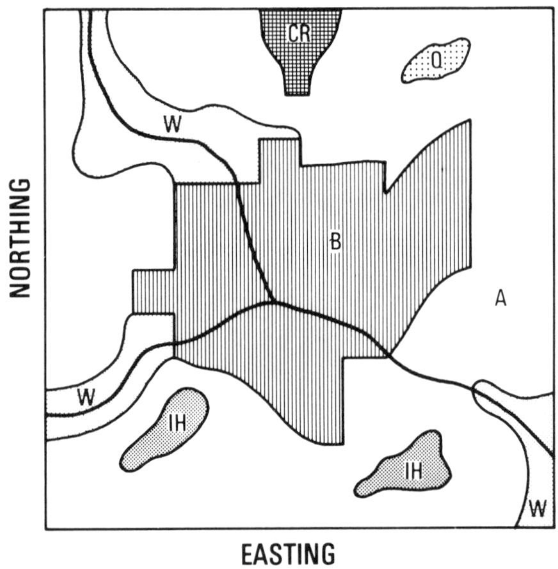

Figure 6.1 A sample land classification.

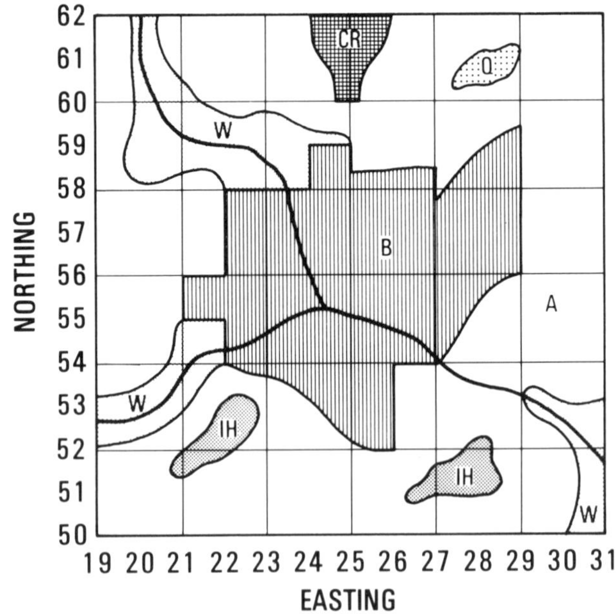

Figure 6.2 Unit areas and land classification.

Table 6.1 Sample Land Use Codes

Code	Meaning
A	Agriculture
B	Built-up Area
CR	Commercial in Rural Area
IH	Intensive Horticulture
Q	Quarry Operations (sand and gravel)
W	Woodlot

SOME METHODS FOR CODING UNIT AREAS

Manual Interpretation

This method is based on an appropriately scaled grid structure of unit areas prepared on a transparent mylar overlay. The overlay is placed on the source map (as in Figure 6.2), and the data classification for each unit area is performed by a clerk. It is usually only possible with moderate accuracy to estimate the largest or predominant single coding for each unit area. Where the unit area is equally divided between two different data areas, the clerk makes an arbitrary assignment. Table 6.2 gives the result of one encoding and an equally correct alternate for the unit area records associated with Figure 6.2. In the table, a unit area is identified, as is common practice, by the geographic coordinates of the point at its center (*i.e.*, its centroid).

In the sample manual interpretation of this simplified example there were four unit areas (comprising more than 10% of the total area) where no clear interpretation was possible. However, in usual practice, only the one code as selected by a research assistant is coded. The codings are written down on data coding forms as interpreted. Later they are keypunched and added to a data file. The last process can add error depending upon the quality of the data entry staff. Errors during the interpretation phase increase with research assistant fatigue. Consistency is lacking as a result of the obvious requirement to employ large teams and to switch staff from task to task. This process is time-consuming and very expensive in terms of labor input. It can only achieve limited accuracy. In spite of these limitations some large data banks have been assembled; one is described in McDaniel *et al.* (1970).

Hardware Remote Sensing Techniques

Remote sensing equipment has been developed which digitizes picture images. This could provide some possibilities for coding unit areas from data maps. A picture is divided into an arbitrary number of unit areas

Table 6.2 A Sample Manual Interpretation

Easting	Northing	Land Use Code	Alternate
20	51	A	
20	53	W	
20	55	A	
20	57	A	
20	59	W	
20	61	W	
22	51	A	
22	53	A	IH
22	55	B	
22	57	A	B
22	59	W	
22	61	A	
24	51	A	
24	53	B	A
24	55	B	
24	57	B	
24	59	W	
24	61	A	
26	51	A	
26	53	A	B
26	55	B	
26	57	B	
26	59	A	
26	61	A	
28	51	A	
28	53	A	
28	55	A	
28	57	B	
28	59	A	
28	61	A	
30	51	A	
30	53	A	
30	55	A	
30	57	A	
30	59	A	
30	61	A	

called "pixels" and the "grey scale" value of each unit area is recorded. The grey scale measures the degree of darkness between white and black on an arbitrary scale. The precision or resolution of state-of-the-art equipment is 256 grey scale levels of a picture matrix of 5 x 7 inches divided into as many as 7.5 million unit areas. The equipment produces records consisting of an x coordinate, a y coordinate and an integer corresponding to the grey scale for each unit area. Computer programs are applied to the data output by the sensing equipment to aggregate pixels to appropriate scale.

A number of difficulties are associated with attempting to use this technique to code arbitrary classification maps. First, only grey scale data can be detected. This requires all area information on a map to be shaded by study team staff prior to analysis by the equipment. Shade legends must be developed and documented to permit data to be used later. Second, the equipment currently available works with the small picture sizes indicated above. This requires source documents to be scaled down using photographic techniques. For example a 24 x 32 inch source map must be reduced to a 5 x 7 negative. Any extreme photoreduction results in significant lens distortion which introduces errors in the data produced. Third, it is necessary to obtain comparable results from analysis of different maps of the same series. It is difficult to maintain exact grey scale from one map to the next particularly when photographic reduction is involved. Difficult photoreduction time consuming recalibration is required.

The earth inventory data currently available from remote sensing agencies utilizing survey satellites is not yet at a quality which is useful in most regional planning applications. Lindgren (1975) reports that LANDSAT1 resolves only at 100 meters and that very little computer software is available to work with the data produced. As a form of data collection for planning purposes he says "I do not foresee remote sensing techniques taking the place of traditional methods." Steiner (1972) agrees; he claims 30-meter resolution is required for traditional regional planning analysis. Lindgren indicates it is often unlikely that either the proposed Earth Observation Satellite (EOS) or LANDSAT3 will resolve at this scale. It is clear that more development is required before these methods will provide data directly useful in regional planning.

Canada Land Inventory Scanning Equipment

International Business Machines has developed a map-reading system under contract from the Canadian Ministry of the Environment. Details of its systems design may be found in Jankulak (1967). The essential characteristics of the data acquisition process may be outlined as a scribing or outlining step, automated scanning, and computer scan interpolation. Under control of an operator, a mylar work sheet is scribed by a machine with which the operator traces the outline of data areas to be input to the data bank. After this step is complete, the scribed sheet is scanned using custom equipment developed by IBM.

The scanning equipment writes a magnetic tape file consisting of grey scale values for unit areas under scan. A large file consisting of 7 million bytes of data is produced for each 30 x 30 inch map. A computer program performs aggregations to identify the nature of data zones. Jankulak describes data-packing techniques used to reduce memory demands but observes that the computer must have at least 512,000 (byte) memory to process the digitized data.

This system was implemented in support of a project to build a detailed data bank of planning information on a regional scale. The initial plans called for data acquisition at the real scale of 1:50,000 for all major population areas and planning areas. This has been revised subsequently to an input scale of 1:250,000 (or one fifth the detail) in large part due to the time required to scribe and scan the maps. The system is based on the resources of a large computer, requires expensive custom equipment and highly trained staff. Its inability to effectively process input at the 1:50,000 scale compromises its usefulness at anything other than general planning exercises over very large study areas.

REPRESENTING OF DATA AREAS BY POLYGONS

Figure 6.3 shows a partitioning of space into six nonoverlapping and exhaustive data zones. Similar to most data zones resulting from land classification exercises, the sample data zones are of varying size and are

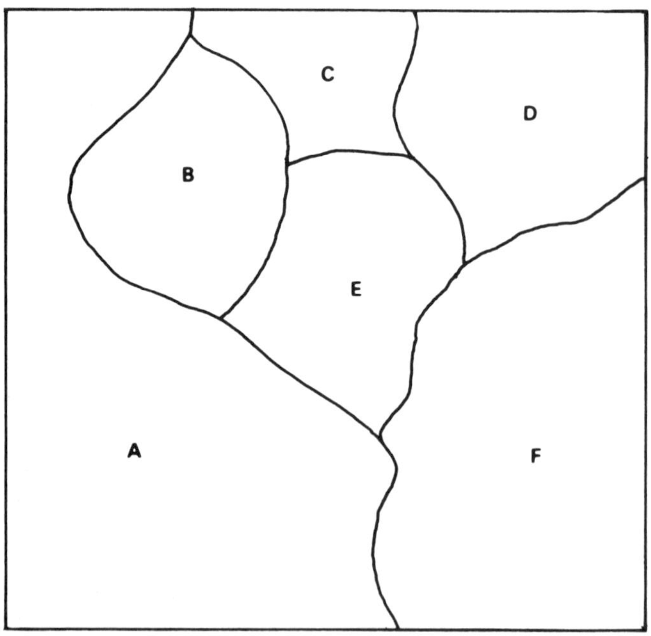

Figure 6.3 Sample data zones.

of nonuniform shape. Popek and Kingston (1972), Peuker (undated), Goodchild (1973a) and Freeman (1974) report upon various methods for recording and storing this spatial information. However, as Freeman summarizes, due to the substantial irregularities in area boundaries "by necessity shapes must be bounded (and described) by a series of straight line segments." A polygon contains the area delimited by the edges which join its vertices. Accordingly each data zone in Figure 6.3 can be thought of as a polygon. A polygon structure of the example is shown in Figure 6.4.

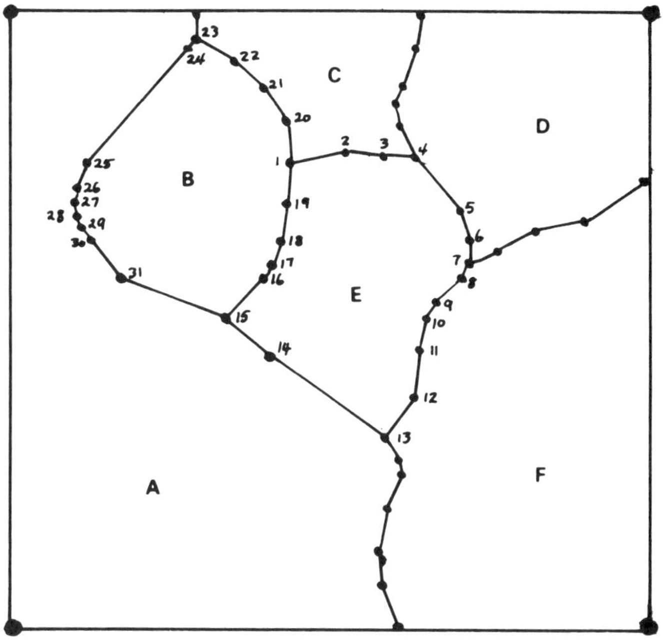

Figure 6.4 A polygon approximation.

Data zone E is approximated by polygon E which has the vertex set:

$$Ve = \{1,2, \ldots, 19\}$$

Similarly, data zone B is approximated by polygon B which has the vertex set:

$$Vb = \{1,15,16,17, \ldots, 31\}$$

Thus, the spatial extent of arbitrary data zones can be recorded by the set of vertices which enclose the area. Vertices are recorded in strict clockwise or counterclockwise order. It should be clear from the examples that the accuracy of the polygon representation is dependent upon the number of line segments used to bound the polygon and the number of line segments used where the boundary of the data zone is changing direction rapidly. Data zones B and E share a common boundary which consists in the polygon approximation of $Ve \cap Vb = \{1,15,16,17,18,19\}$. Data zones A,C,D,F are considered to be polygons with the outside boundary of the sample area as part of their polygon outline. Unit areas can be thought of as a series of simple four-sided polygons. Thus, the unit area data coding problem can be considered as determining the nature of the intersection between an arbitrary unit area polygon and the set of data zone polygons. Such an intersection set is shown in Figure 6.5.

COMPUTER INTERPOLATION OF POLYGON-BASED UNIT AREA CODINGS

Area Intersection

Goodchild (1973b) has developed computer programs which algebraicly determine the set of polygons which result from intersecting unit area

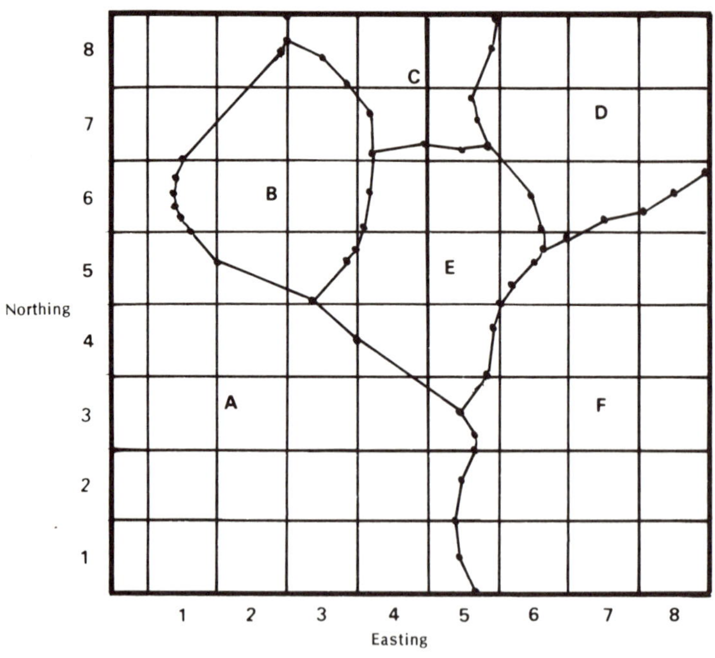

Figure 6.5 Polygon-unit area intersection.

polygons with data area polygons. In other words, individual data area polygons are fractured into smaller polygons which are contained in unit area polygons. The resulting data set consists of many polygons, the largest of which is a unit area and the smallest of which is a partial unit area in size. The process is complex. It involves, in part, determining the intersection vertices between all data areas and unit area polygons and then assembling the appropriate polygon descriptions. The resulting vertex set consists of all unit area vertices, all data vertices and all intersection point vertices. This represents a substantial increase in data storage requirements over the representation of the polygons by unit areas. Maps involving many complex data areas require substantial computing to derive the intersection polygons.

Once all fragmented intersection polygons have been determined, each of the intersection polygons is identified according to the unit area in which it lies. The unit area identification is used to develop an aggregated record for each unit area. This method of determining unit area characteristics was rejected because it is not yet well developed and appears to be restricted to moderate-sized, moderate-complexity source documents.

A Sampling Method

In Chapter 3 fundamental accuracy limits were mentioned for the standard 1:50,000 map series. The implication is that very accurate unit area and coding analysis is not consistent with the accuracy of the mapped data. Accordingly, the author has developed an interpolation system based on a sampling approach which requires moderate resource input in terms of special equipment, staff and computer time. The method is based on using a point to represent a small area around the point. In other words, when the nature of the unit area can be determined at a point, then the nature of the immediate area around the point can be estimated.

Polygons are nonoverlapping and exhaust the study area. With the exception of points on polygon boundaries, any arbitrary point selected within the study area will be contained in one polygon. Points on boundaries are arbitrarily assigned to the polygon having the closest center. Determining the containing polygon is possible by applying an appropriate "point in polygon" algorithm. An arbitrary number of such data points can be selected and evaluated for each unit area. The larger the number of points per unit area, the smaller the area whose nature is being estimated by a single point and, accordingly, accuracy is increased. Figure 6.6 shows the results of three different interpolation sampling densities on

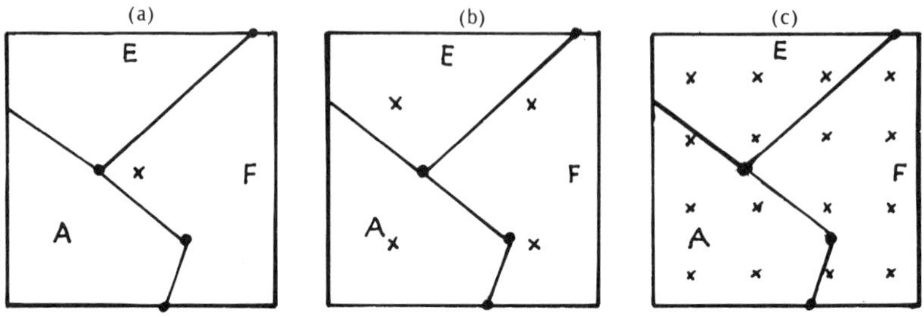

1 Point Interpolation		4 Point Interpolation		16 Point Interpolation	
A	0 %	A	25 %	A	31 %
E	0 %	E	25 %	E	25 %
F	100 %	F	50 %	F	44 %

Figure 6.6 Point in polygon interpolation.

unit area (5,3) from Figure 6.5. The results of the 16-point interpolation are very close to the exact results for A,E and F of 34; 26 and 40%.

A computer program based on a sampling approach was implemented to automate data gathering; it is described in a following section.

Digitizing Data Areas

This step establishes the coordinates of a series of polygons which define all data areas on a map. The source document is mounted on a drafting table which is equipped to measure with great accuracy, the (x,y) position of a cursor. An operator moves the cursor around each polygon outline in a continuous counterclockwise or clockwise direction causing a series of digital (x,y) records to be generated which describe vertex coordinates relative to an arbitrarily selected origin. The task can be easily performed by clerical staff.

For each data area, a series of punched data records are generated indicating the classification code for the area followed by the set of points defining the area. When all data areas have been processed in this way, a polygon data set is available for unit area interpolation. The specification of the geocode or geographic coordinates of the origin used for the map sheet and the scale of map are all that are required to permit the interpolation program to interpolate the data to the appropriate unit areas.

Certifying Data Area Accuracy

It is essential to assure the accuracy of the polygon representation of the basic data zones to obtain an accurate set of unit area evaluations. The data entry procedure involves a detailed examination of the digitized polygons. This is facilitated by utilizing a computer-controlled plotter to draw on translucent paper the exact polygon representations at the same scale as the source document. This permits a clerk to overlay the polygon representation directly over the source document and locate boundary inconsistencies. Figure 6.7 shows the data collection and certification process. Polygon outlines are corrected, overlay maps redrawn and checked on the light table iteratively until the polygon structure accurately represents the source document data areas. As a result, the data representations input to the computer should be as accurate as on the source documents.

The verified polygons, map scale and geographic coordinates for the map sheet origin are input to the interpolation program along with directives which specify the sampling density to use. The program overlays a unit area grid on the polygon structure, samples the specified number of points in each unit area, and develops a frequency count for each unit area of unique polygon characteristic codes detected. Depending upon parameters set by the user all unique codes are recorded along with their

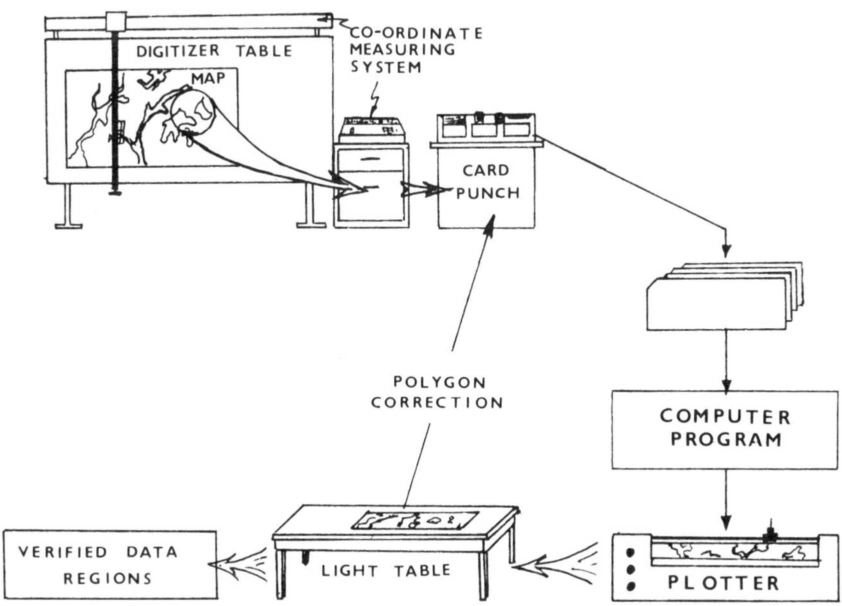

Figure 6.7 Polygon digitizing and certification.

percentage frequency (*i.e.,* area) in a computer disk file. The user may request that only a maximum number of unique polygon codes be considered; in such a case those with the largest percentage area are recorded.

Various sampling densities are available for selection. For a given utility planning study, the assessment team may choose to use different sampling rates for different map series. This would be done to match source document precision and recording accuracy required for impact factor assessment. Sampling patterns are pre-set to provide a uniformly consistent coverage across all unit areas. The appropriate sampling rate is selected by the assessment team based upon the size of the minimum feature to be detected. The selection is based upon the tabulated accuracy results and computational implications found in Chapter 7. Usually a pilot interpolation run is performed for a complex data area and the results evaluated prior to finalizing the sampling rate. Once the rate is fixed, source maps for a data series are interpolated one at a time as shown in Figure 6.8.

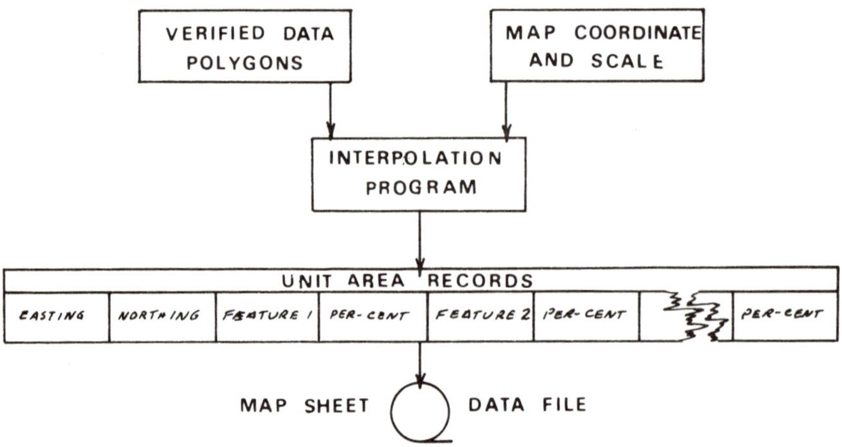

Figure 6.8 Unit area character interpolation.

CHAPTER 7

DATA INTERPOLATION BY SAMPLING

DATA AND INTERPOLATION SCALE

As outlined briefly in Chapter 6, each data zone on a source document is represented by a polygon. The actual data describing a polygon consists of a series of points (x_i, y_i) for i = 1, . . . , n listed in a continuous tracing order where x_i and y_i are the number of one hundredths of an inch in the x or y direction the point is from an arbitrary origin. The user specifies delta "D" the number of one hundredths of an inch there are in one side of a unit area increment of a coordinate and can specify the actual geographic coordinates (Rx, Ry) of the digitizing origin for a source document. It was found useful to provide a delta X "Dx" and delta Y "Dy" value for different precisions in the X and Y coordinate directions. Certain manuscript maps which have been duplicated a number of times have been found to stretch in one direction as much as 8%. Some published production map series have been found to have X and Y scales out by as much as 5%. The conversion of polygon vertices to geocode is:

$$GX_i = \frac{X_i}{D_x} + R_x$$

$$GY_i = \frac{Y_i}{D_y} + R_y$$

The actual number of data zones and their boundary irregularities vary greatly from map to map and map series to map series. Simple 1:50,000 topographic planning maps may have as few as 50 polygons with an average of 10 to 12 vertices per polygon; a very detailed land use map may contain 1500 polygons with an average of 8 vertices per polygon.

The scale for unit area size can be set independently from the real polygon scale on the document. Since unit areas are of fixed square size, the entire set of unit areas to be interpolated can be specified by giving the least X, least Y, greatest X and greatest Y values of the equivalent source area along with the length of one side of a unit area in geographic coordinates. This means that source information can be supplied at any arbitrary scale and the unit areas still interpolated at a different desired scale. This removes the necessity to redraft nonstandard map series.

INTERPOLATION OUTLINE AND POINT IN POLYGON ALGORITHM

The interpolation program involves input of data area polygon structures, their conversion to geographic coordinates and storage for repeated reference. The interpolation program examines the unit areas on the source document from north to south and east to west. For each unit area the appropriate sample points are generated and checked for membership against the stored polygon structure. As the examination of each unit area is completed a descriptive record is written to secondary (disk) storage. Due to the irregularities of polygon shape and size it is necessary to check each possible polygon for possible point membership. This requires an accurate and efficient point in polygon algorithm and effective data storage to accommodate the polygon descriptions.

Numerous point in polygon algorithms are known; they are outlined in Stinton *et al.* (1972) and evaluated for accuracy in Aldred (undated). They agree that the algorithm designated the "topological algorithm" is the only general algorithm which gives consistently accurate results. This algorithm was selected for the interpolation program. It is based on reference line and polygon boundary intersection and may be described as follows:

a) Given P a polygon where P is described by a set of straight lines $P = \{L_1, L_2, \ldots, L_n\}$
b) Given (x,y) a point
c) To determine if (x,y) lies in P then extend a sufficiently long line (called the reference line) in one direction from (x,y) and count the number, I, of $L_i \in P$ such that L_i intersects the reference line.
d) If I is even the point (x,y) is not contained in P. If I is odd the point is contained in P.

The algorithm encounters two degenerate conditions in a general application. The first is the problem of intersecting a vertex or a series of vertices giving potentially wrong counts. The second involves intersecting a parallel line giving I = ∞. However, both of these problems can be overcome by detecting the conditions, changing the reference line, and reapplying the algorithm.

Merrill (1973) has used a similar basic algorithm in a special case and improved speed by partitioning coordinates into Y classes having increasing X values. This permitted him to perform a quick exclusion test before applying the full algorithm to each polygon. While his revised algorithm was not directly applicable, the author has successfully applied, as described below, the principle at two levels in the topological algorithm implementation to realize increases in efficiency.

DATA STRUCTURE AND ALGORITHM IMPLEMENTATION

The topological algorithm is based upon line intersections. This requires the representation of polygon structure as a series of line segments. Three basic representations for line segments are possible: (a) representation by two points; (b) representation by one point and a vector length and angle; and (c) representation by the point-slope form and extreme values of x and y associated with the line.

Representations (a) and (b) require identical amounts of computer storage and two-thirds of the storage required for (c). However, (a) requires the calculations of the slope and intercept each time the line is checked for intersection with another. Method (b) requires trigonometric function evaluations to determine line intersections. While representation (c) requires more storage, it requires little computation to determine the point of intersection and additional information is available which can speed up the topological algorithm. Accordingly, representation (c) was chosen to represent the polygons.

Since each polygon has an arbitrary number of sides, it is appropriate to store the line segments in a list structure. In addition a polygon identification list containing the addresses of the appropriate entries in a line segment list for each polygon is defined.

ALGORITHM SPEED-UP MODIFICATIONS

In order to by-pass full application of the topological algorithm for each polygon, use was made of the least orthogonal containing rectangle to by-pass redundant calculations.

Definition 7.1: The *least orthogonal containing rectangle* is a rectangle whose sides are parallel to the coordinate axis, contains a specified line or polygon and is the least area rectangle satisfying these conditions.|

Examples of least orthogonal containing rectangles are shown in Figure 7.1. It is evident from the figures that such a rectangle can be completely

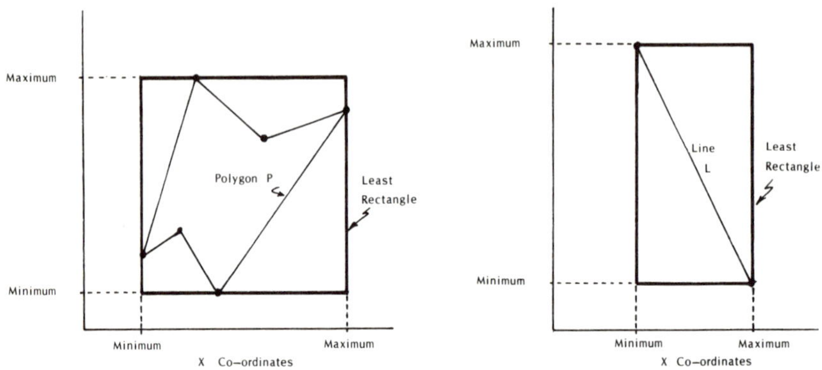

Figure 7.1 Least orthogonal containing rectangles.

described by four values associated with the shape: minimum x; maximum x, minimum y and maximum y coordinates. It is further evident that for a point (x,y) to lie in polygon P, it must also lie within the least orthogonal containing rectangle for P. For two lines to intersect, they must have a common point. For any point (x,y) to be on a line L, then that point (x,y) must lie in the least orthogonal rectangle containing L. Accordingly, for lines L_1 and L_2 to intersect, their intersection point must be contained in the least orthogonal rectangles. It is also clear that if the least rectangles do not intersect that there cannot be an intersection between lines L_1 and L_2.

Each polygon identification list entry includes the coordinate values defining the least rectangle containing that polygon. The addition of this least rectangle information permits the partitioning of the set of all polygons with respect to an arbitrary point (x_i,y_i) into the subset whose least rectangle contains (x_i,y_i) and the subset whose least rectangle does not. The containment test for least rectangles with respect to a point (x_i,y_i) is achieved in only four comparisons. This means that the topological algorithm need only be applied to polygons where there is a possibility of the point being contained therein. This partitioning is possible without accessing the line segment list.

Additional speed-up is achieved in the actual topological algorithm by utilizing least orthogonal rectangle partitioning on a line segment basis. Once it has been determined that point (x_i,y_i) lies in the least rectangle of polygon P, it is necessary to attempt to intersect each of the polygon's line segments with the reference line drawn in one direction from (x_i,y_i) and count the intersections. By maintaining a least orthogonal rectangle for the reference line and each of the line segments, it is possible to partition the set of line segments of polygon P to the subset whose least rectangles overlap with that of the reference line and the subset whose do not. It is then only necessary to perform exact intersection analysis for the first subset. The resulting data structure is shown in Figure 7.2.

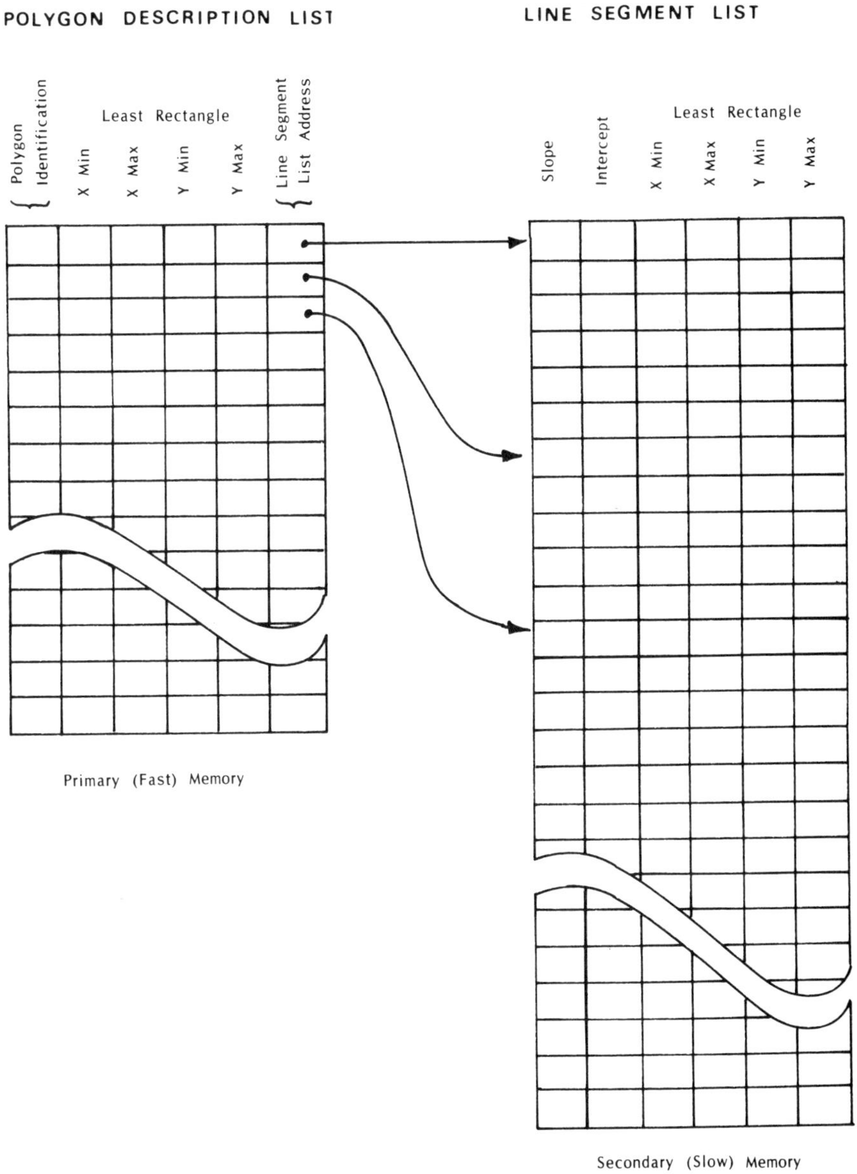

Figure 7.2 Polygon data structure.

During program execution, the polygon identification list is steadily accessed; it is held in fast central memory. The larger and less active line segment list is held in secondary memory. This helps to reduce computing costs.

SAMPLING DENSITIES

To permit the user to control the accuracy at which the interpolation is to proceed, the program contains options to apply, for any one of a set, a predefined sampling pattern in all unit areas. Figure 7.3 shows the

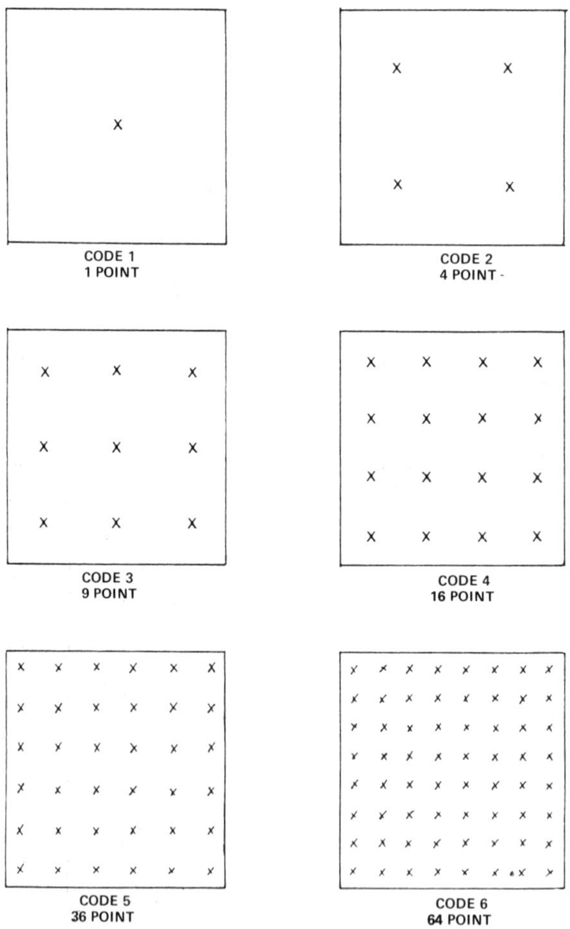

Figure 7.3 Six sampling patterns.

first six sampling patterns. It will be observed that the patterns have been developed to maintain a uniform sampling rate across the entire study area. Table 7.1 shows the sampling densities which are available. Sample patterns are recursively produced by generating four- or nine-point patterns around four- or nine-point sample cluster centers. Computer time required is linear to the number of sample points per unit area; however, the number of sample points goes up rapidly with increased option codes.

Table 7.1 Sampling Accuracy

Basic Accuracy		
Interpolation Code	Number Points	Average Smallest Area Detectable (%)
1	1	100.0
2	4	25.2
3	9	11.2
4	16	6.0
5	36	2.8
6	64	1.6
7	81	1.2
8	144	0.72
9	256	0.38

Accuracy Relative to 500-Meter Unit Areas			
		Required Resolution	
Interpolation Code	Smallest Area Unit Area	Site Resolution (Meters)	For 1:50,000 map (Inches)
1	61.8	500	0.40
2	15.6	251	0.20
3	6.9	167	0.13
4	3.7	122	0.10
5	1,7	82	0.066
6	0.98	63	0.050
7	0.74	55	0.044
8	0.44	42	0.034
9	0.23	30	0.024

SOME CONSIDERATIONS

It is evident that the accuracy of unit area interpolation will increase as the sample density is increased although it will be seen in Figure 7.4 that the increase is not linear. The appropriate sampling rate is dependent upon the size of the smallest item required to be detected. An additional consideration is the basic accuracy of the source document; it makes no sense to interpolate at an accuracy level finer than is available in the

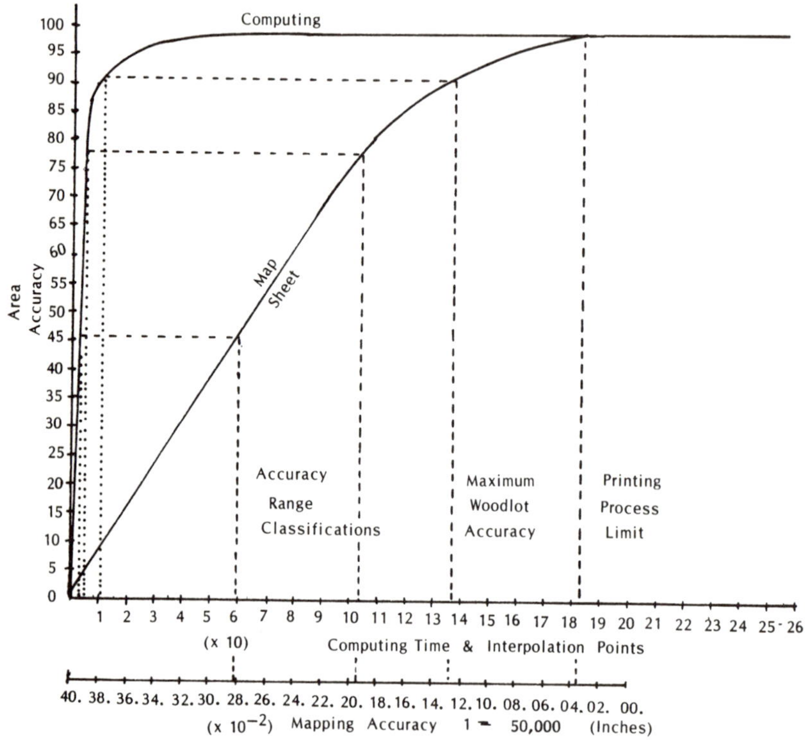

Figure 7.4 Sampling accuracy and computing.

source document. Table 7.1 shows the area accuracy of the sampling technique independent of the size of the unit area. In addition it relates the accuracy to a 500-meter unit area and the 1:50,000 topographic survey. These results are summarized in Figure 7.4 where it will be observed that increased accuracy beyond the 5% level is achieved only at greatly increased computing time. This is because computing time increases by 2x when X is the interpolation code.

In the figure, the accuracy limit for mapped data at 1:50,000 is indicated. If the accuracy limit is referenced to interpolation level, it indicates, for example, that nine-point interpolation is nearly as accurate and sixteen-point interpolation exceeds the accuracy of woodlot plotting on the 1:50,000 maps. Further inspection will show that interpolation at four or nine points is usually adequate for most mapped classification studies mapped at 1:50,000 and analyzed on a 500-meter unit area basis.

CHAPTER 8

PHASE II: IMPACT FACTOR EVALUATION

In any application of the planning system a number of separate impact factors may be developed. As previously indicated, an impact factor is a mapping from a subset of the study data base to an impact ranking.

IMPACT FACTOR MAPPINGS

Domain of the Mapping

An impact mapping U_i for factor "i" usually operates on variables belonging to the unit areas which are being evaluated. The rating Z for a unit area is obtained by:

$$Z = U_i\ (X_a, X_b, X_c, \ldots, X_f)$$

The mapping may be extended to include in its domain special functions on data base variables for the specific unit area being mapped or variables in the immediate neighborhood of the unit area. The latter mapping is called a "neighborhood function" by Peuker and Chrisman (undated). If G is a mapping on a unit area's variables and N is a mapping on a set of variables belonging to the unit areas which are its neighbors then the general mapping for a unit area is:

$$Z = U_i \left\{ X_a, X_b, \ldots, X_f, G(X_k, \ldots, X_n), N(X_p, \ldots, X_t) \right\}$$

when $\{a, b, \ldots, z\}$ is an appropriate indexing set.

Range of the Mappings and Impact Ranking

The planning system is designed to be flexible; it permits impact mapping ranges to be the set of real numbers. However, in most environmental assessment problems, only a ranking of the impact associated with a given unit area is required. In this context, it is usual to define the range of the mapping as a subset of the set of positive integers. For example, $Z \in \{1, 2, \ldots, r\}$ where $r \geqslant 2$ is the maximum ranking possible for an impact factor. While it is possible that each factor could map to a different subset of Z, it is advisable that the same subset of the integers be used for each factor. This facilitates comparison between impacts and and their later composition into a single composite impact. It is common practice to assign the ordinal values 1,2,3, . . . , r to indicate impact intensity ranging from none to heaviest.

Fisher and Davies (1973) comment that threshold impact conditions should be defined. This is supported by McTaggart-Cowan (1973), Malisz (1972) and Rattray *et al.* (1973). The threshold condition is one where irreversible or very extreme impact conditions are detected. Such a condition would be ranked "r". McTaggart-Cowan proposes four classes or ranks of impact as shown in Table 8.1.

Table 8.1 Sample Impact Rankings

Rank	Description	Meaning
1	Compatible	Immediate recovery or no impact at all; no environmental management required.
2	Moderate	Impact exists–recovery will take place in time; no environmental management is required.
3	Severe	Impact exists–no recovery is possible without environmental management.
4	Threshold	Impact exists which will never recover even with environmental management.

McTaggert-Cowan's rankings involve a combination of both the concepts of recovery and management. However, they do not include adequately temporal change. He provides for only three impact ranks below threshold; this appears to be very restrictive. This lack of definition could give rise to large extents of code 2 impact. The corresponding lack of resolution will provide reduced guidance for selecting corridors for utility development. The author has extended McTaggart-Cowan's rankings as shown in Table 8.2.

This set of rankings provides for increased resolution at the cost of adding very little complexity. Final rankings used in a planning system application are selected by the assessment team and can be as simple or complex as desired. The requirement to provide maps and tabulations for nonspecialist review during external review should mitigate against overly

Table 8.2 General Impact Rankings

Rank	Description	Meaning
1	None	No measurable direct or indirect impact or slight impact which is immediately recoverable not requiring environmental management. No significant change forced in use or quality of the area.
2	Slight	A temporary light direct or indirect impact or minor forced change in use or quality or area. Complete recovery takes place in a year or less without significant management.
3	Moderate	A temporary direct or indirect impact or forced change in use or quality of the area. Seventy-five percent recovery in two to five years and complete recovery in ten years with no significant management required.
4	Heavy	A mainly temporary direct or indirect impact or forced change in use or quality of the area. Fifty percent recovery in five years and seventy-five percent recovery in ten years with some management required.
5	Severe	A long-term direct or indirect impact or any major forced change in use or quality of the area in a negative direction. Fifty percent recovery in ten years with careful management.
6	Prohibitive	A permanent loss of a significant use or quality of the area. No recovery possible.

complex rankings. Peuker (1972) claims that the mapped results of five class rankings can be relatively well perceived by specialists and non-specialists alike while seven class rankings begin to strain even the specialist's perception.

IMPACT FACTOR MODELS

The planning system is oriented to using special purpose subroutines in an impact analysis program to evaluate individual impact factor mappings. This permits the assessment team to develop as simple or complex an assessment algorithm as required for each individual impact factor mapping. A number of general methods are outlined briefly. In practice, an impact factor algorithm may be a combination of any number of these methods.

Linear and Nonlinear Summation Models

Simple linear models have been the basis of most attempts to perform impact analysis on a unit area basis. This is evident in the work by

Rattray *et al.* (1973), Krauskopf and Bunde (1972) and Lyle and Von Wodtke (1974). These models require at least ordinal and preferably interval data values. However, most environmental data consist of ordinal and nominal classifications. Subsidiary mappings are defined on such data to provide the appropriate interval scaled values for combination purposes. If $\{G_1, G_2, \ldots, G_n\}$ is a set of such subsidiary mappings and $a_1, a_2, \ldots a_m$ a set of coefficients specified by the assessment team then the basic model for linear impact mapping is:

$$Z = \sum_{i=1}^{n} a_i\, G_i\,(X_{\Delta i}) + \sum_{i=n+1}^{m} a_i\, X_{\Delta i}\, ; \; m > o,\; m \geqslant n,\; n \geqslant o$$

The model gives each data base variable (*i.e.*, G_i and X_i) relatively equal importance (save their weightings by the set of coefficients $\{a_1, a_2, \ldots, a_m\}$). As a result, a number of low values can reduce the effect of one high value (and the reverse) as the sum for a particular unit area is being calculated. In general, values for Z can tend to cluster around the mean values for each variable scaled by its coefficient in the mapping. This results in an impact factor assessment with relatively little resolution to assist in corridor selection.

Nonlinear models may be used to give extra prominence to specific data base variables in a mapping. In this case, the model for Z becomes:

$$Z = \sum_{i=1}^{n} a_i\, G_i\,(X_{\Delta i})^{\beta_{\lambda i}} + \sum_{i=n+1}^{m} a_i\, X_{\Delta i}^{\beta_{\lambda i}}\; ; \; m > o,\; m \geqslant n,\; n \geqslant o$$

where Δi and λi are members of the appropriate indexing sets and at least one $\beta_{\lambda i} \neq 1.0$. However, this model can also produce values which cluster around the mean of means of each coefficient times its powered variable. While some variables will be given more or less dominance in the sum due to their power coefficients, the impact factor assessment is likely to provide little more resolution than the simple linear model.

Both linear and nonlinear summation models share the additional problem that values of Z are not likely to provide simple rankings (*i.e.*, $1,2,3, \ldots, r$). This requires the addition of a classification function. Such functions are described in a following section. Including the classification function, the general summation mapping model becomes:

$$Z = j \text{ when } C_{j-1} \leqslant \left(\sum_{i=1}^{n} a_i\, G_i\,(X_{\Delta i})^{\beta_{\lambda i}} + \sum_{i=n+1}^{m} a_i\, X_{\Delta i}^{\beta_{\lambda i}} \right) < C_j\; ;$$

$$C_j = C_j + \delta^{\gamma_{\mu j}} \quad j = 1, \ldots, p$$

where C_o, $\{\delta_1, \delta_2, \ldots, \delta_p\}$, the indexing set $\mu_1, \mu_2, \ldots, \mu_p)$ and p are defined according to the discussion in the section on classification mapping.

Classification Mapping

A classification mapping can serve as an impact factor mapping on a variable or as a method of remapping the range of a separate impact factor mapping. It essentially maps a value to an integer rank or class number of p (arbitrary) possible classes. While many *ad hoc* classification mappings are possible the most general may be defined as follows:

$$C = i \text{ when } C_{i-1} \leqslant z < C_i\,;\; C_i = C_{i-1} + \delta_i^{\beta \Delta i} \text{ for } i = 1,2, \ldots, \ell$$

when C_o, the set $\{\delta_1, \delta_2, \ldots, \delta_p\}$, the set $\{\beta_1, \beta_2, \ldots, \beta_p\}$ and the appropriate indexing set for Δ_i are defined by the assessment team. In simple cases the class increment δ_i and C_o may be constant and defined relative to p, and the maximum possible range of values of Z as follows:

$$C = i \text{ when } C_{i-1} \leqslant z < C_i\,;\; C_i = C_{i-1} + \frac{Y_{max} - Y_{min}}{\ell}\,;$$

$$C_o = Y_{min} \text{ for } i = 1,2, \ldots, \ell$$

This appears to be one of the most popular classification methods and is called the *equal interval classification* mapping. While this mapping is easily specified and implemented, it has the disadvantage that one cannot guarantee that the frequency of values in classes developed by C will have any uniformity. Indeed, it cannot be guaranteed that all classes will have nonzero frequencies. When this classification mapping is applied to a variable Z which is Nor (μ,σ), selecting class number ℓ can have the effect that

$$k\,\sigma \cong \frac{Y_{max} - Y_{min}}{\ell}\,;\; k \in J \text{ the integers}$$

The implications with respect to class frequencies are summarized in Table 8.3.

Mapping Rankings

Data base variable values often consist of rankings or a series of ordinal classifications. It may be desired to map a number of these variables to a single rank for an impact factor. In addition, it is possible to formulate

Table 8.3 Class Frequencies When the Class Interval is An Integral Multiple of the Deviation

Multiple of Deviation (k)	Number of Classes Containing Over 95% of all Data	Comments
1	4	Largest two classes will contain roughly 34% of observations each.
2	2	The largest class could contain as much as 68% of the observations.
3	2	The largest class could contain as much as 86% of the observations.
4	1	This multiple of the deviation and over forces all observation into one class.

an impact factor algorithm which develops a number of distinct classifications using, for example, summation and classification models and then combine the classifications to provide the impact factor assessment. Since the variables or classifications being mapped are not interval in nature, summations and averages are not appropriate. Two traditional methods are described here. Other methods include the new Cascade Algorithm and the matrix method described in later sections.

Dominant Frequencies

A frequency table can be developed according to the data values observed in a unit area. The value with the maximum frequency is chosen as the result of the mapping. This method relies on each variable having a near identical range in values. The method completely ignores the *ordinal* nature of the variable values. For example, the addition of one high or one low value is unlikely to alter the mapping unless it would cause the respective frequency to become maximum over all other frequencies. This may be inappropriate in assessing an impact factor since an especially high or low ranking may be very significant.

Maximum (Minimum) Threshold

The value assigned for this mapping is the maximum (or minimum) value of all its variables for a unit area. This counteracts the problem identified in the dominant frequency approach but it ignores entirely the frequency at which specific rankings were observed. In practice, it can be convenient to develop *ad hoc* methods which combine both the dominant frequency and maximum frequency methods.

Mapping Rankings or Nominal Values by Matrix Methods

In cases similar to those outlined previously and in cases where nominal data are involved it may be necessary to develop a different mapping. Since each variable's values consist of a finite subset of the integers, they usually can be used as an index on one dimension of a classification or decision matrix which contains the appropriate value for the mapping. For example, suppose "soil type" was recorded as four classes and "drainage" was recorded as three. A six-class mapping is defined by the contents of the table as shown in Figure 8.1.

	SOIL TYPES				
	Class	1	2	3	4
	1	1	2	4	4
DRAINAGE	2	2	3	5	6
	3	3	5	6	6

Figure 8.1 A matrix classification mapping.

Such matrix mappings need not be limited to two-dimensional matrices. Within reason, additional dimensions can be added to accommodate the number of variables required. It is not necessary to limit the variables only to those of nominal or ordinal values. Variables with interval values can be accommodated if their number of distinct values is small or their values are passed through an appropriate classification mapping. The entries in the table are arbitrarily assigned by the assessment team. This may be a straightforward task for matrices of low dimensionality but the task becomes difficult as dimensionality increases. For example a matrix for eight variables, each of which has four classes (or values), requires 128 matrix entries to be specified.

TWO OTHER IMPACT FACTOR MODELS

Diversity Assessment

At times it is difficult to render an assessment in useful dimensionless units. When establishing an index of relative value of areas for use for outdoor education, Scott and Nelson (1973) developed a diversity index to overcome this problem. By doing so they avoided some of the difficulties of assigning weights directly to esthetic judgments. They assemble a list of items or elements which, if present, are perceived to add to the value of an area. The diversity index is simply a summing of all the number of items in an area which match list elements. They assume that areas have high value if their diversity sum is large and low value if their sum is small. Their conjecture is supported by Vaughn (1973) who claims that complex vistas and landscapes have more scenic value than the less complex. Diversity assessment could prove very useful in providing part of a visual or scenic impact assessment during a utility routing study.

Local Surface and Neighborhood Analysis

Cut-fill operations are significant in the planning of highways, pipelines and railways. The assessment of a specific unit area in this context is not only a function of its own elevation and slope, but also its relative elevation and the slope to the immediate neighboring surface.

Local surface analysis can be used to help evaluate and cut-fill operations. Figure 8.2 shows a cross section view of a cut-fill area. Figure 8.3 shows how the fit of a plane to local surface permits the calculation of the residual from local surface at unit area. Each unit area is located by X and Y geographic coordinates and has an elevation which can be denoted by Z. A local surface fit is performed for a bounded region around a unit area at (x_i,y_i) by estimating the coefficients for the elevation function "e" such that

$$Z = e(x,y)$$

for some arbitrary order of e. (In the figure a plane or first-order fit has been performed.) This can be achieved by using a standard least-squares method. A residual elevation r is defined as

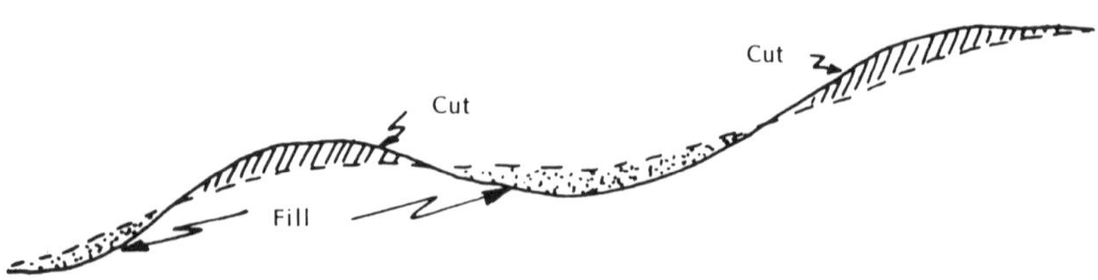

Figure 8.2 Cut and fill cross section.

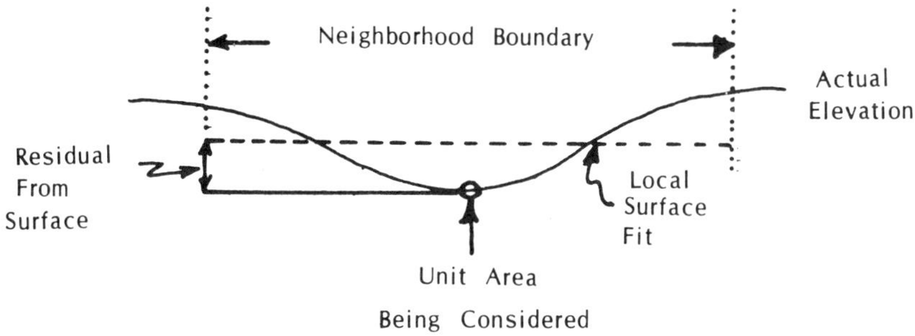

Figure 8.3 A local surface fit.

$$r = a_{ij} - e(x_i, y_i)$$

where a_{ij} is the actual elevation at point (x_i,y_i) and $e(x_i,y_i)$ is the elevation estimated by the local surface fit at point (x_i,y_i). If r is negative, a fill operation is required; and if r is positive, a cut operation is required. The magnitude of r is proportional to the amount of material effected at that unit area. A series of such local surface fits can be performed as shown in Figure 8.4.

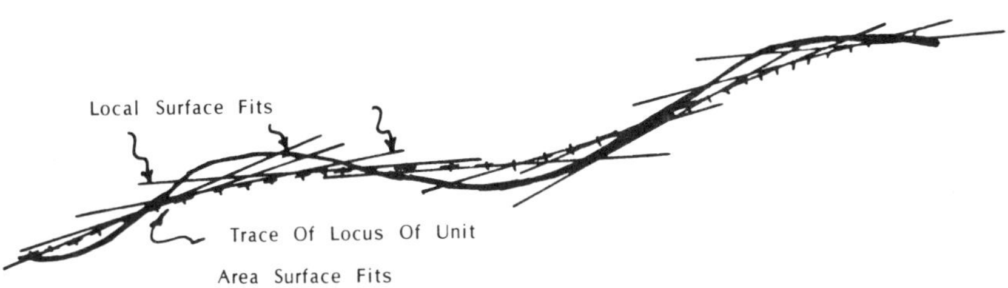

Figure 8.4 Cut fill by local sequential surface fits.

As a corridor is followed from some starting unit area S to terminal unit area T, the integral:

$$\int_S^T \left| a_{ij} - e(x_i,y_i) \right| d_i$$

gives the amount of material moved in cut-and-fill operations. An appropriate-sized neighborhood is arbitrarily selected to assure that gradients are held within specifications.

Local surface fitting can be accomplished by accessing neighboring unit area data values as the impact factor algorithm is being applied to a unit area or by precalculating the surface fits and storing the value of the surface elevation along with actual elevation during the development of the data base.

EVALUATION AND ADJUSTMENT OF IMPACT FACTOR MODELS

Assessment team members develop impact factor models through a series of iterations. At a later phase the results of impact factor assessments are subjected to external review. This may require further refinements by the assessment team and subsequent iterations. During the development of the factor models, subject area specialists select test areas for pilot study evaluations. The impact factor algorithm is applied to the area and mapped for review. Direct field comparisons are made, modifications made to the model, impact re-mapped and field comparisons repeated until the desired results are obtained. The model is applied to the entire study area after it is found satisfactory. The results of its application are mapped and should be subjected to the review of subject area specialists outside the assessment team.

Explicit consideration of present versus future impact can be included in an impact factor. A useful way to proceed is to develop a model of present impact and a separate model of future impact based upon the appropriate data base information. Indirect impacts can be modeled by a separate model. These separate models can be combined by an explicit weighting formula which can be tailored to reflect emerging public and governmental concerns. Separate submodels are useful since they can be developed, assessed and certified independently. When it becomes evident that an adjustment of the weights between present, indirect and future impacts is required, only the coefficients need be altered.

CHAPTER 9

PHASE III: COMPOSITE IMPACT ANALYSIS

Phase III provides the synthesis of the results of phases I and II. It combines, on a unit area basis, the individual impact factor assessments into a single composite ranking. In phase II, the impact factors themselves were assessed on unit area basis yielding impact ranks $Y_1, Y_2, Y_3, \ldots, Y_k$ for the k different factors. Composite impact rating C is also derived for $F = \{Y_i \mid i = 1, \ldots, k\}$ on a unit area basis. This maintains maximum localization of impact assessment and is essential for effective routing analysis in phase IV.

Since impact factors are developed independently, they should provide dissimilar impact rankings for individual unit areas. For example, an impact factor defined for agricultural operations will likely rate a rocky outwash stream valley area as low impact. A natural environment and recreation factor might rank the impact for the same area as heavy or severe. Similarly, there may be divergent assessments considering urban areas from the point of view of different factors. In some studies it is possible that all factors will not be completely unanimous in their impact ranking for any one unit area.

To maintain consistency with the development of impact factor ratings, it is important that the prohibitive impact rating (described in Table 8.2) for any individual impact factor be carried into the composite impact rating. In other words, if a unit area is assessed prohibitive impact or "no development" by any one impact factor, the unit area should attain a final no development status independent of impact among any other factors.

Fisher and Davies (1973) observe that the culmination of moderate impacts gives rise to a threshold of impact. They observe that threshold exists for each local environment (*i.e.*, unit area) with respect to its capacity to sustain change; they state "once these thresholds are exceeded by cumulative developments, disproportionate cumulative environmental changes are induced and the environment is permanently altered." It is important that the combination of impact factor rankings reflects this fact. The established methods do not handle cumulative impacts properly; the new Cascade Algorithm developed by the author is a solution.

ESTABLISHED METHODS

Weighted Average

This is the method of composition used by Krauskopf and Bunde (1972) and Lyle and Von Wodtke (1974). The model applied is

$$C = \frac{\sum_{i=1}^{k} a_i Y_i}{\sum_{i=1}^{k} a_i} \quad \text{for arbitrary weight set } W = \{a_1, a_2, \ldots, a_k\}$$

It is important to note that a few low-valued impact ratings can disguise a very high prohibitive rating. For example, assume eight impact factors had been ranked $\{1,3,1,4,2,1,3,6\}$ for a specific unit area and the model being used was equal weighted (*i.e.*, $W = \{a_i \mid a_i = 1, i = 1, \ldots, 8\}$). Then the composite impact is 2.63. This characteristic of the linear summation model violates the concept of prohibitive threshold impact and has lead Dickert (1974) to observe "a single index conceals impacts."

It is natural in such circumstances to attempt to alter the weighting set W to emphasize favored impact factors. This, however, does not solve the problem. For example, if the weighting set $\{1,1,1,1,1,1,1,3\}$ was applied to the unit area rankings previously given, the resulting composite is only 3.3. In addition, as soon as one variable is weighted heavily relative to others the consistency in meaning between individual impact factor ratings is compromised.

Under uniform weighting, the clustering of individual composite impact values for C around the mean of the means for the impact factor ratings will tend to give a composite assessment with extensive areas of similar rating. This lack of resolution reduces the effectiveness of the route analysis phase. Graphic description of this phenomenon may be seen in the weighted average composite map in Chapter 11.

The appropriateness of the weighted average combination model must be addressed when the values of the Y_i impact factor rankings are ordinal. In the preceding examples, two "composite ranks" were derived: 2.63 and 3.3. Their meanings are unclear. Is the first to be considered a two- or a three-rank impact? To be correct in using a linear summation model, a function I (X) should be developed to convert the ordinal rankings to interval values before calculating the average. In addition the resulting composite rank should be mapped by a classification function to derive the final composite ranking.

Threshold Rankings

This method is applied to provide a partial resolution of the objection (above) raised by Dickert. An additional advantage is that it is appropriate to apply this method to ordinal values as well as interval data; no ancillary function I (X) is required. The method is very simple; it simply selects for the composite impact ranking the maximum ranking observed:

$$C = \max \left\{ Y_1, Y_2, \ldots, Y_k \right\}$$

This method of composite impact assessment provides an improved impact resolution. This may be observed in the threshold composite map in Chapter 11. Prohibitive and high impact areas are identified irrespective of the factor defining them. However, concerns about assessing cumulative impacts are not adequately addressed by this method. For example, the number of times a particular rank is observed does not have any effect on the model.

NEW CASCADE ALGORITHM

General Description

The algorithm was developed to produce a composite assessment which permits identification of sensitive areas which result either from high impact measured by a single high-valued impact or a number of moderate impacts measured by some subset of moderate-valued impacts. The algorithm does not require interval data values. It is applied directly to the ordinal impact ratings. The essential feature of this algorithm is the promotion of multiple lower-level impact ratings to higher-level impact ratings. The promotion effect is dependent upon the rating level where promotion is permitted to begin in an upward direction. It is also dependent upon the number of identical ratings which must be combined to produce one next higher-level rating.

The algorithm is applied on a unit area basis. It begins at an arbitrarily selected low-impact level called the cascade level, no ratings below this level are considered as candidates for promotion. The algorithm steps up level by level each time locating the number of impact ratings at that level. An arbitrary number of identical ratings required for a single promotion to the next level (called the cascade base) is used to determine the number of promotions at each level.

Impact factor rankings to be promoted can be considered as leaf nodes in a tree graph where levels of the tree (*i.e.,* distance out from the root) represent ranking values. For example, suppose there were seven leaf nodes with values; they could be written down in vector form (1,1,1,2,1,1,1). Let it be decided that any two like values should be "promoted" and represented as the next value in sequence. The effect can be demonstrated by recursively applying a "rewriting rule" to the vector, the rewriting rule being "rewrite any two like values as the next level value always using the lowest value pair." The results are shown in Figure 9.1.

Initial Vector	**(1,1,1,2,1,1,1)**
	(2,1,2,1,1,1)
	(2,2,2,1,1)
	(2,2,2,2)
	(3,2,2)
	(3,3)
Final Vector	**(4)**

Figure 9.1 Upward promotion by vector rewrite.

This rewriting chain can be represented in a tree diagram as shown in Figure 9.2 where the level of the root determines the final value of the composite ranking. In the diagram, the original nodes are denoted □;the nodes which result from promotion are denoted as ○.

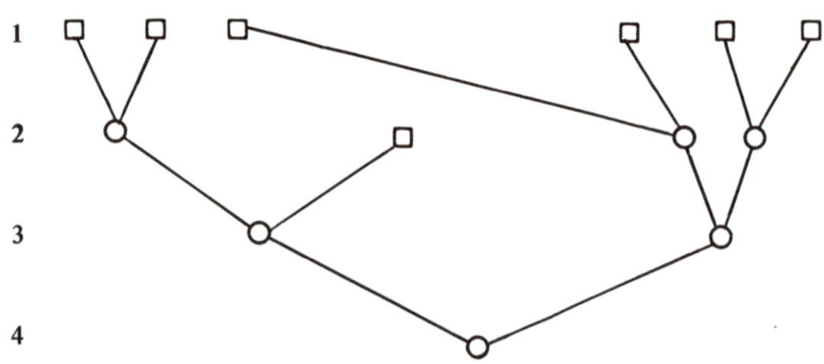

Figure 9.2 A promotion tree.

The figure should make clear the fact that the amount of promotion which could take place is dependent upon the number of data values, the values themselves, the number of like values required for promotion, and the level (or minimum data value) at and above which nodes will be considered candidates for promotion.

The assessment team arbitrarily specifies the number of like values which are required to force a promotion. This number is called the cascade base *b* in the algorithm. A matter of interest is the maximum promotion effect relative to the arbitrary base and number of original data values. Clearly, maximum effect will take place when all original variables have the same value. The estimation problem may be restated as: given a set of leaves or nodes on a tree at a given level and the number of branchings at a node, how far is the root from the leaves? The solution is related to *b*, the cascade base, and *k*, the number of leaves. When one considers that from the root there are *b* edges giving rise to *b* nodes and for each such *b* node, *b* edges giving rise to yet another set of nodes etc., it becomes evident that the number of levels, n, has the following property:

$$\text{n is maximum integer such that } b^n \leqslant k$$

To solve for n:

let $\quad b^n = k$

then $\quad n \cdot \ln b = \ln k$

thence $\quad n = \dfrac{\ln k}{\ln b}$

However, there are only integral number of levels in a tree structure and the solution becomes

$$n = \left[\frac{\ln k}{\ln b} \right]$$

where [X] denotes the greatest integer of X. For example, consider the promotion tree in Figure 9.3 where the cascade base is 3.

Now $\quad \ln k = \ln 10 = 2.302581$

$\quad \ln b = \ln 3 = 0.77424$

$\quad x = \ln 10/\ln 3 = 2.9$

and $\quad n = [x] = 2$ the number of promotion levels in Figure 9.3.

(for k = 10, b = 3)

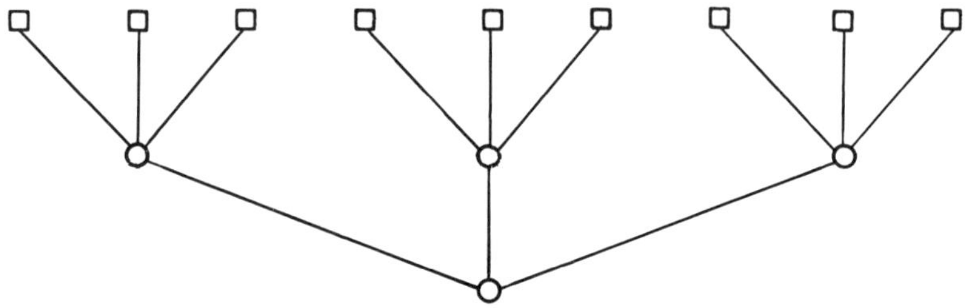

Figure 9.3 A maximum promotion.

Statement of Algorithm

The algorithm proceeds as follows:

Let $r' = \max \{Y_1, Y_2, \ldots, Y_k\}$ over all unit areas.
Let $r = \max \{Y_1, Y_2, \ldots, Y_k\}$ for the specific unit area.
Let b, a positive integer, denote the arbitrarily selected cascade base.
Let $\#(F)$ represent the order of a set F.
Define $A_i \leqslant \{F | Y_j = i; j = 1, \ldots, k\}$ for $i = 1, \ldots, r$ and
$A_i = \phi$ for $i > r$.
Set q, the maximum possible rating which can be obtained from the application of the algorithm, as $$q = r' + \left\lceil \frac{ln\ k}{ln\ b} \right\rceil$$

Denote by L the arbitrary cascade level.

Algorithm steps are:

1. If $r < L$, then cascade rating is r; exit from algorithm.
2. Set $i = L$.
3. Set $n_i = 0$.
4. Set $n_i + 1 = \left\lceil \dfrac{n_i + \#(A_i)}{b} \right\rceil$
5. Set $n_i = n_i + \#(A_i)$.
6. (Optimizing step) if $i > r$ and $n_{i+1} = 0$, then do step 8.
7. If $i = q$, do step 8; otherwise continue at step 4 with $i = i + 1$.
8. Then C is the cascade rating when $C = \max i$ such that $n_i \neq 0$ and $i \leqslant q$.

Application and Comparison

The algorithm as stated in the preceding section will provide numerical values greater than the maximum or prohibitive impact ranking when a number of severe or prohibitive impact data values are encountered. In such a case, the value for C is arbitrarily set to the code for prohibitive impact since it is the highest impact rating. The algorithm has been successful at detecting high impact areas which are a combination of moderate impacts and provides good resolution.

The result of a sample application of the algorithm is shown in the series of Maps 1 through 6. The first map is an enlargement of a portion of a 1:50,000 topographic survey for a sample study area. Maps 2 through 5 show four independent impact factor assessments for the same area. Map 6 shows the composite map which resulted from an application of the algorithm where cascade base b and level l were each assigned the value 2. The darkest shades on the maps are prohibitive impact or "No Go" areas; the lightest shades indicate slight impact areas. One should note that the resolution on the composite map has facilitated locating alternate corridors between points A and B. A visual comparison of the factor maps will show that different factors give quite different assessments of impact for identical areas. This is particularly true near the town of Simcoe at the middle of the sample region. In the composite map, locations in the area near Simcoe have promoted impact ratings due to the algorithm's cascading effects.

An example of algorithm application to the values in a single unit area may be compared to the weighted average and threshold ranking methods. Let the cascade base and level be set at 2 and let the weighting matrix be defined $W = \{1,1,1,1,1,1,1,1\}$ for the weighted average case. Consider now the application of the algorithm to eight factor ratings:

$$1,3,1,4,2,1,3,3$$

The algorithm is applied starting at level 2; however, there is only one two-level rating. The first promotion will then come by pairing the two threes. Figure 9.4 shows the resulting cascade promotion chain.

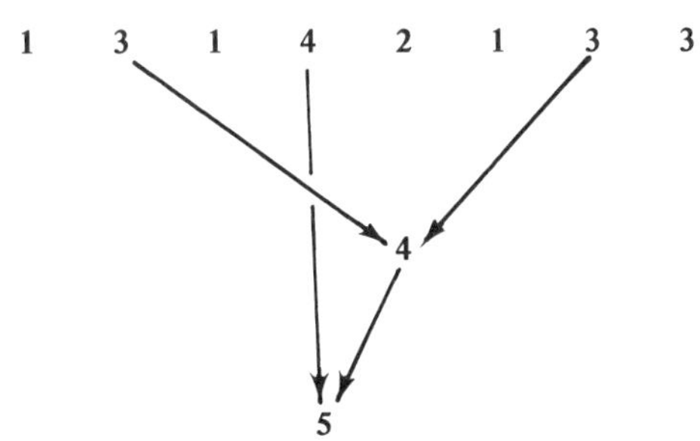

Figure 9.4 A sample cascade promotion.

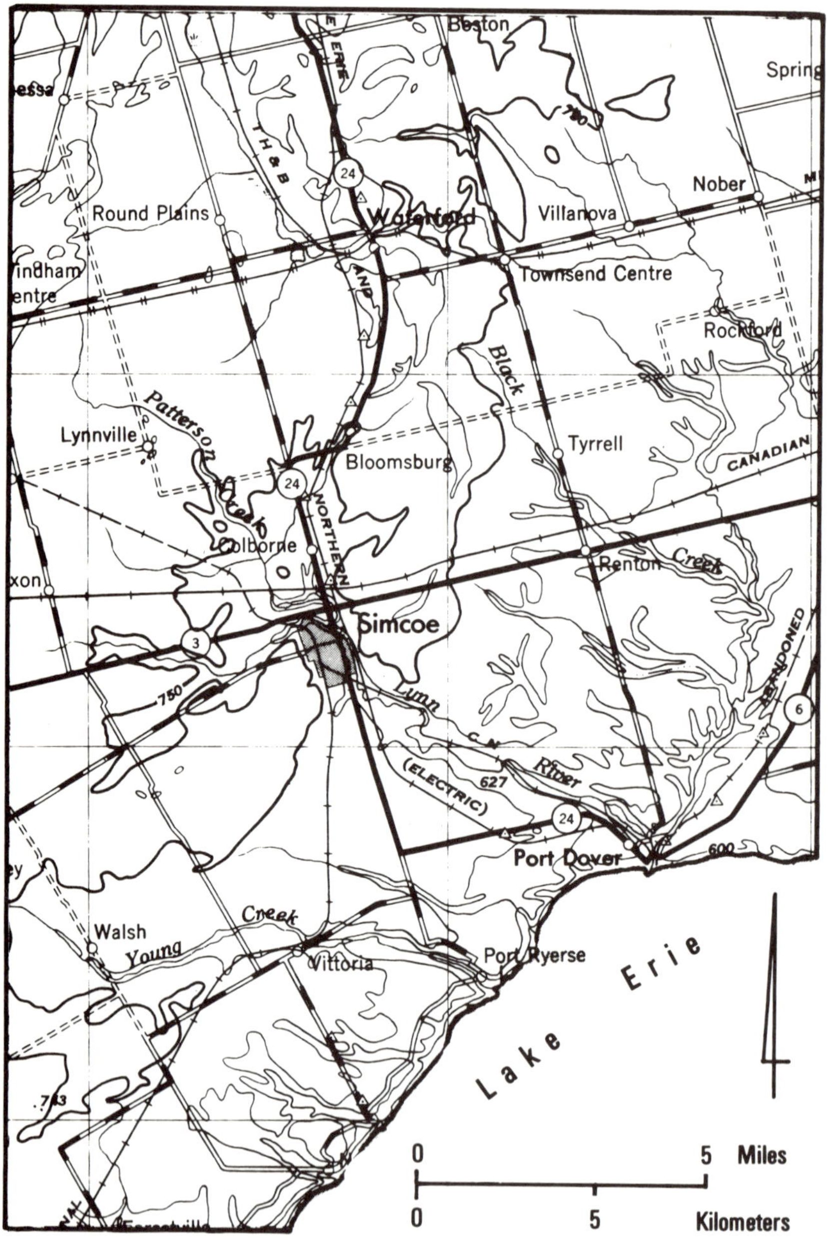

Map 1. Study segment, topographic survey.

Map 2. Agricultural operations factor.

Map 3. Natural environment factor.

Map 4. Recreation, cultural and historical factor.

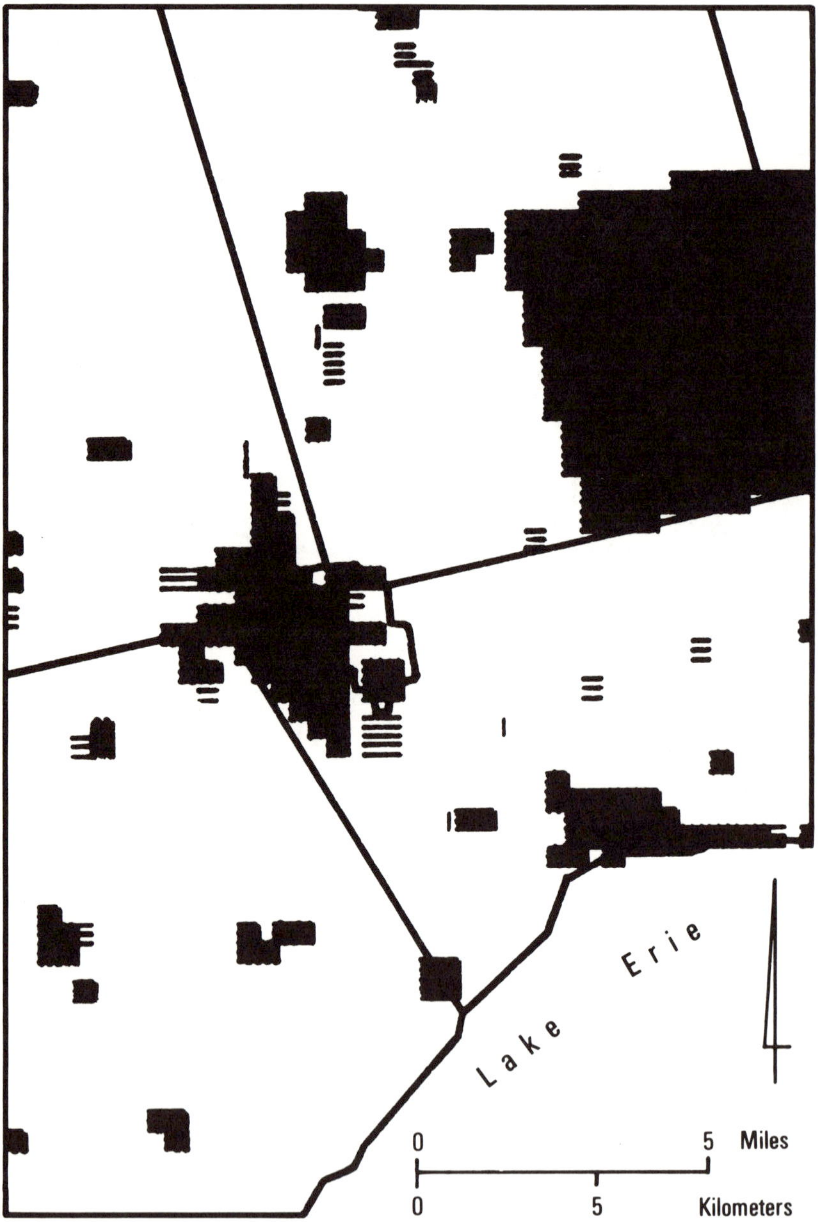

Map 5. Residential, institutional and commercial.

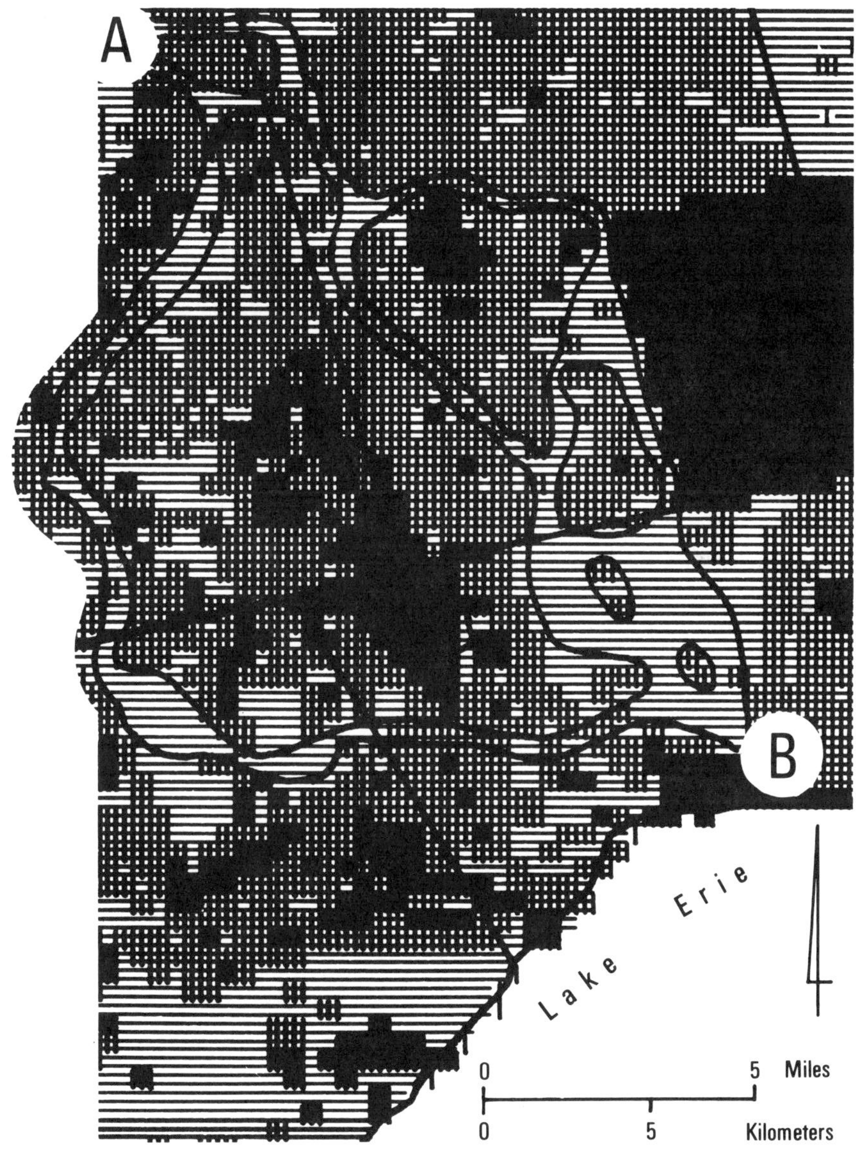

Map 6. Cascade algorithm composite of maps 2, 3, 4, 5.

Thus, the composite rating for this unit is five. The equal weight average yields a composite rating of 2.25 and the threshold rating is 4. In actual studies one can anticipate that a larger number of impact factors will be used. The study team may wish to try different applications of the algorithm using different cascade base and level parameters to obtain the appropriate resolution. Of course, it is possible to develop subcomposite maps using fewer than the full set of impact factors. This could be done to develop, for sectors which are represented by more than one impact factor, mappings for analysis and presentation purposes.

POSSIBLE APPLICATIONS

The cascade algorithm may find application where "voting" assessment is required. Areas for potential application are telemetry signal and digital picture processing. The algorithm might be used to give a single composite from a number of signal channels. Scene analysis of digital pictures is facilitated if there is good resolution. In this case the algorithm might be used incrementally across a moving "frame" of picture elements to assess the value for a central picture element. This should serve to increase picture resolution. In remote sensing, satellite transmissions consist of a series of picture images on a number of separate bands. The algorithm might be used to provide a single synthesis of the information being received on all bands.

CHAPTER 10

PHASE IV: ROUTE DEVELOPMENT AND ANALYSIS

CONSIDERATIONS AND GENERAL DESCRIPTION

At the end of phase III, the planning system has developed a composite impact surface for the entire study area. The surface itself can be expected to be very discontinuous as isolated pockets of high impact show up in areas of low impact, and the converse. The problem is now to locate corridors for utility development between two terminal points arbitrarily defined by the project proponent. Corridors need to be developed so that they are "least cost" in terms of environmental impact. Route finding must be performed to the level of accuracy available in the composite impact surface. The existing methods for route finding are discussed in Chapter 2 and their associated difficulties are outlined. A new method is now described which develops alternate corridors.

SOME ROUTE DEVELOPMENT RELATIONS TO GRAPHS

The method described in this chapter is based on a graph representation of the problem. In the following discussion, numerous references are made to graph theory concepts and definitions provided in Chapter 2. A sample arrangement of unit areas is shown in Figure 10.1(A); associated with each of the unit areas is its data value. This grid-organized collection of unit areas is a graph where each unit area is delimited by the edges joining the four vertices which are located at its corners. The planar dual graph to the graph in Figure 10.1(A) is shown in Figure 10.1(B) where each unit area is now represented as a vertex and the edges connecting vertices represent the connection relationships between unit areas. In the dual graph the data value of a unit area is associated with its

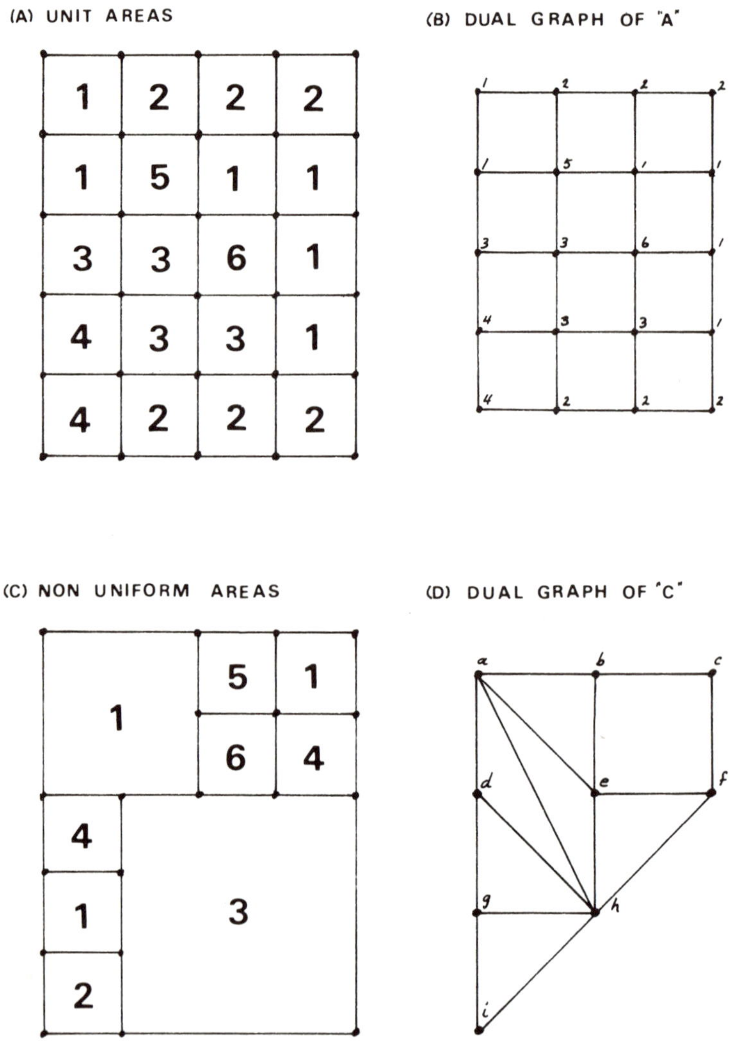

Figure 10.1 Unit area graph and its dual.

corresponding vertex. In the example, distances between adjacent unit areas are identical since the areas are uniform squares. Accordingly, this dual graph not only records the connection relationship between unit areas but intercell distances as well. Dual graph vertices may be assigned arbitrary identification codes if required.

When areas of nonuniform size are involved, as shown in Figure 10.1(C), the dual graph [see Figure 10.1(D)] still maintains the connection relationships. Interarea distances, however, are not constant in graph 10.1(C) and are represented in the dual graph by assigning "weights" to each edge. Note that arbitrary vertex letters have been used. The weights corresponding to the dual graph are shown in Table 10.1 where edges are identified by the two vertex numbers which they connect. They represent the distance from the center of each area to the center of its neighbor.

Table 10.1 Edge Weights for Dual Graph "C"

Edge	Weight
(a,b)	1.97
(a,e)	1.97
(a,h)	2.92
(a,d)	1.97
(b,c)	1.00
(b,e)	1.00
(c,f)	1.00
(d,h)	2.23
(d,g)	1.00
(e,f)	1.00
(e,h)	2.00
(f,h)	2.23
(g,h)	2.00
(g,i)	1.00
(h,i)	2.23

Various methods are known for finding shortest paths between vertices in weighted graphs. The major ones have been outlined in Chapter 2. Planning system route development is based upon Dijkstra's method. Four distinct steps are performed by computer programs to develop and assess alternate route corridors. They involve, first, graph contraction and adjacency weightings; second, alternate corridor development based on a modified Dijkstra algorithm; third, corridor straightening; and fourth, alternate corridor analysis. Planning system, computer software, data, and study team interaction for this phase are shown in Figure 10.2.

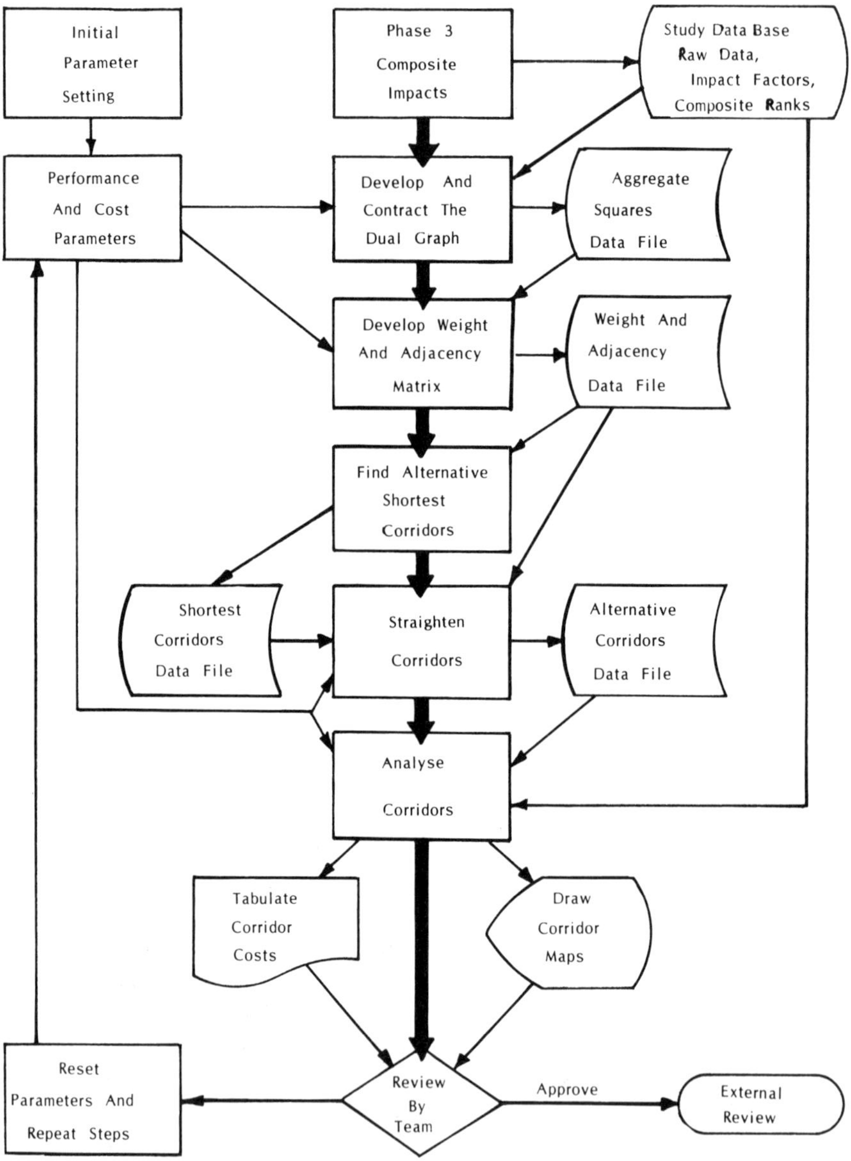

Figure 10.2 Phase IV operations.

GRAPH CONTRACTION AND ADJACENCY WEIGHTINGS

The Dual Graph and Its Contraction

The dual graph representation of the routing solution provides some additional benefits in addition to permitting the use of a (modified) established routing algorithm. Restructuring the dual graph leads to some reductions in data storage and in computer time. This permits large studies to be performed. Provision can be made to guarantee that corridor width is above an arbitrarily defined minimum. This aids the practicality of the resulting study.

The representation of graph (B) of Figure 10.1 requires the recording of 20 vertices and 31 edges. It would be advantageous if the dual graph could be in some way simplified to reduce its size and simultaneously maintain the correct record of connections between areas of common value in the unit area graph. In Figure 10.3 graph (A) replicates on a unit area basis the information shown in Figure 10.1(C). An ordered dual graph can be developed for graph (A) by adding edge direction indicating the "greater than or equal" impact relationship. Such an ordered dual graph is shown as graph (B) in Figure 10.3. If an edge shows a direction from some vertex "a" to some vertex "b" this indicates "b" has a data value (*i.e.*, rank) greater than or equal to that of vertex "a". Where vertex "a" is equal to vertex "b" in data value, directed edges are added in both directions.

The directed dual graph includes not only connection information but ordering information which indicates the relative value of vertices. Graph reduction techniques are applicable to directed graphs where strong components are located and replaced by a single vertex. As indicated by its definition in Chapter 2, a property of a strong component is that it is a minimal non-null (in or out detached) subgraph. This means that all edges connecting the set of vertices to the remainder of the graph must be either indirected or outdirected. Inspection of graph (B) in Figure 10.3 leads to the detection of three components which are designated in graph (C) of the figure as C_1, C_2 and C_3. The strong components are "factored out" of the original dual graph. Each component can be represented by a single vertex, and the ordered relationship between the value of the components can be shown by directed edges between them. The remainder of the graph is left intact as shown in Figure 10.3(D).

The slightly reduced graph (D) can be used to locate more components as indicated. This results in a component structure and reduced original

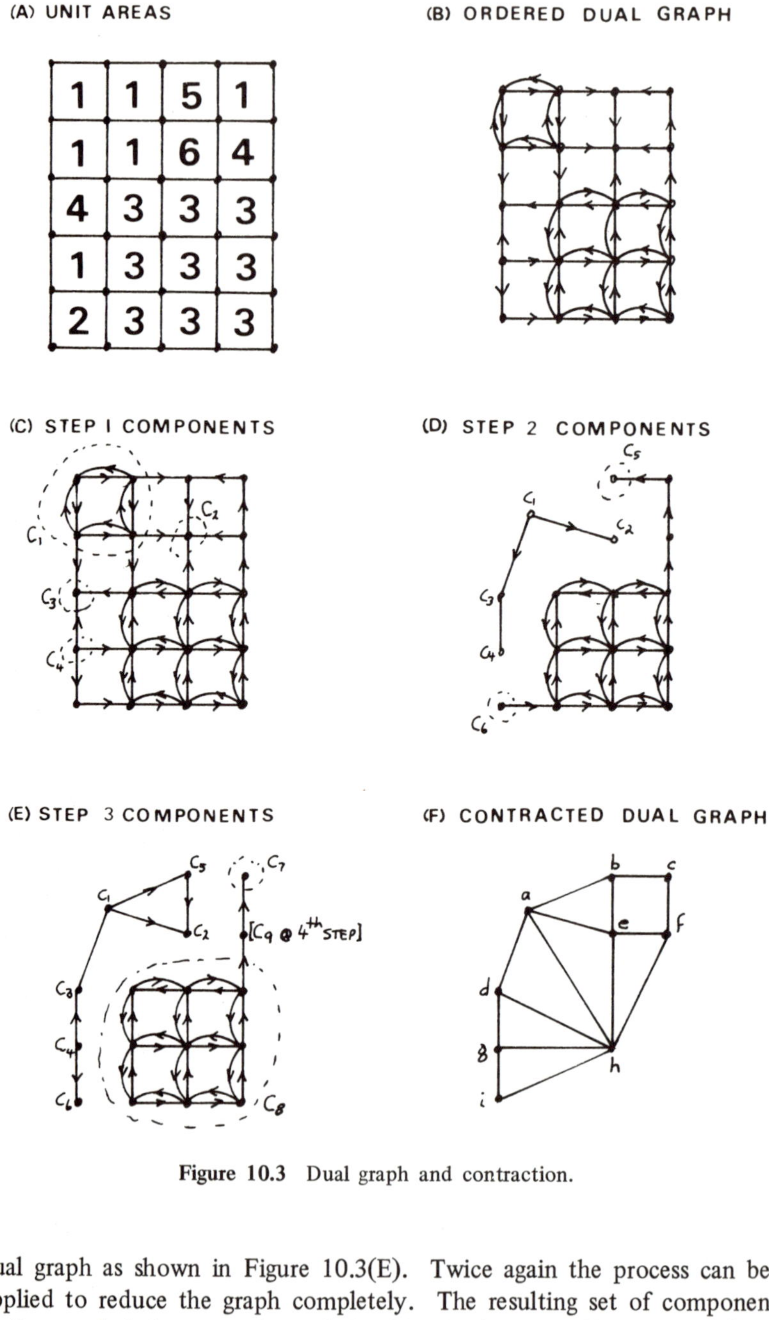

Figure 10.3 Dual graph and contraction.

dual graph as shown in Figure 10.3(E). Twice again the process can be applied to reduce the graph completely. The resulting set of component vertices and their connection relationships is shown in Figure 10.3(F).

The reduced dual graph is the same as the one in Figure 10.1(D). Thus, in the process of detecting graph components, the areas of like nature were isolated and are now represented as vertices. The resulting reduced dual graph gives the essential value-oriented area relationship and may be represented as 9 vertices and 15 edges in comparison to the 20 vertices and 31 edges of the original dual map. The unit area on graph 10.1(D) itself had consisted of 30 vertices and 49 edges.

The resulting reduction in number of vertices and edges required to represent the area relationships reduces computer memory demands and facilitates much larger studies. In addition, the reduction in number of vertices and edges can greatly reduce the computer time required to perform routing analysis. The reduction preserves value discontinuities over space as originally recorded in the unit area graph. Efficiency has been gained by recording with detail (*i.e.*, a number of vertices and edges) areas of maximum discontinuity, while using minimal information (*i.e.*, only a few vertices and edges) to record areas of similarity.

Corridor Continuity and Regular Areas

In establishing corridor routings through an impact surface represented by a contracted dual graph, corridor continuity must be assured. Unrestricted graph contraction can give rise to the development of irregular areas. Data area irregularities make it difficult to determine what the appropriate intervertex distances should be. This is due to the fact that it becomes difficult to determine the center point of the area of complex shapes. For example, in degenerate conditions, it is possible that the area-weighted center of an area is located outside the actual area. In addition, irregular areas make it difficult to assess what configurations of connecting lines are appropriate for expressing intercell distances. These difficulties can be reduced by restricting graph reduction to produce only simple regular areas.

Lemma 10.1

The straight line joining the centers of two adjacent orthogonal squares is contained in the two squares.

Proof:

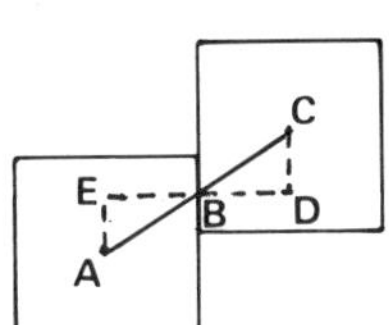

As shown in the diagram:

> Let a square of side length S have as its center A.
> Let C be any point outside the square.
> The straight line AC intersects a side of the square at B.
> Since the length of BE = 0.5 S, then AE $\leqslant$ BE, and AB is in the square.
> Let C be the center of an adjacent square with side length T.
> By the orthogonality of the squares, BD = 0.5 T, CD $\leqslant$ BD and BC is in this second square. Now AB is in the first square and BC is in the second. Thus B is a point on each square's side common to both. Accordingly the line AC is contained in the two squares.

It can be seen by Lemma 10.1 that if only square areas are contracted to single vertices in the dual graph, one can be sure the connection between two such adjacent vertices represents a distance entirely contained in the two areas. This assures that a continuous route exists between the two areas. If a route is assembled from such a set of continuous subroutes, the total route is continuous. Thus, the graph contraction is carried forward in such a way as only to contract to square regions. While this does not bring about complete contraction of the dual graph, significant reduction does take place.

Graph Contraction and the "Square" Vertex List

The process of contracting graph components to squares is performed by a computer program called the "square finder." Input to the program consists of a small set of parameters and the matrix of composite impact ratings derived at the end of phase III of the planning system. The algorithm develops a "square" vertex list for the final set of squares wherein is recorded the X and Y coordinate locating the square, the length of a side and the data value (*i.e.*, impact ranking level) for this (possibly aggregated) area. Parameters may be used, under assessment team control, to relax square finding criteria in later iterations of the planning system. These parameters are described later.

The program finds squares by sequentially contracting all the 1 level areas, then all the 2 level areas, and so forth. For an arbitrary impact level, the impact matrix dual vertices are scanned from north to south and east to west until a vertex having that arbitrary level is detected. Once such a vertex is located, the program attempts to establish the largest possible square contraction with the vertex as the north and westmost part of the square. This is done by performing a simultaneous scan to the east and south until a vertex with a different value is located. Figure 10.4 shows these steps as they affect the source unit area graph

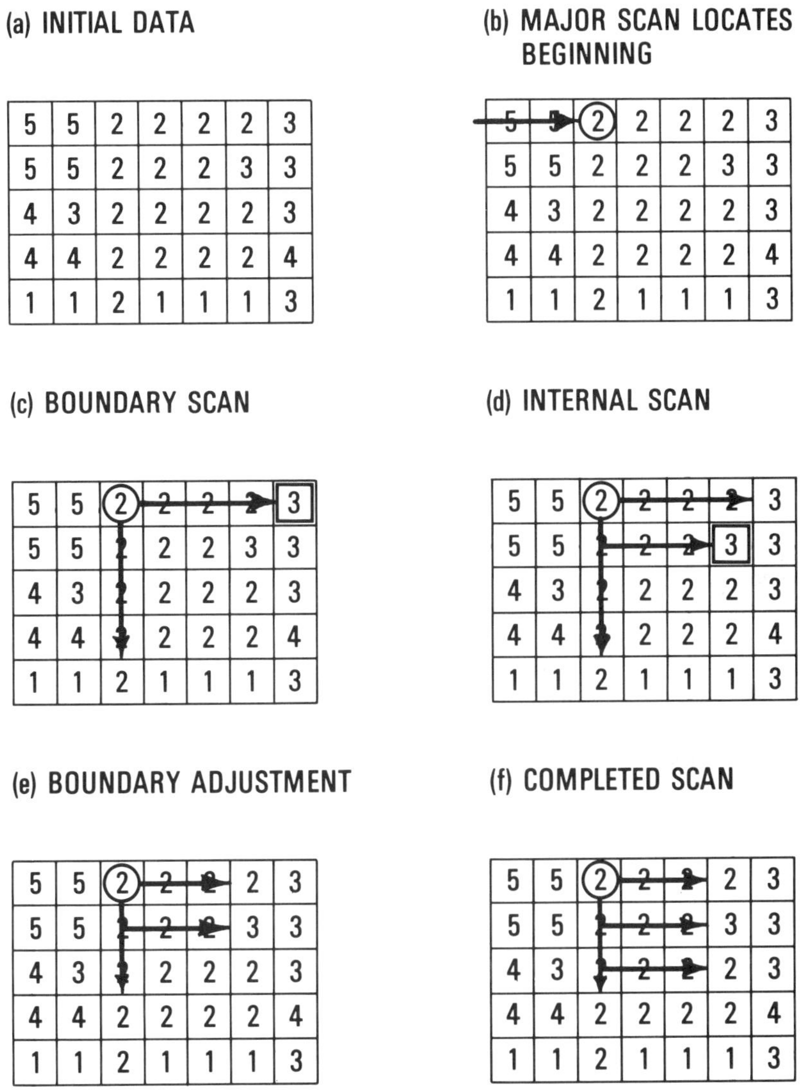

Figure 10.4 Example steps in square finding.

when developing level 2 squares. Part (a) of Figure 10.4 shows the initial data; part (b) shows the scan for 2 level data locating an area to begin contraction. Part (c) shows that a trial boundary for a 4 x 4 square is established resulting from the 3 value in the boxed square being detected when the scan attempted to go to a 5 x 5 square.

Once a trial square boundary is established, an internal scan to assure internal area consistency begins as shown in Figure 10.4(d). When an inconsistency in value is detected as shown in part (d), the trial boundaries are reduced as shown in part (e) to remove the inconsistency. The internal scan is continued until an internally consistent square area is detected as shown in part (f) of Figure 10.4 and each scan of an internally consistent square is completed, a single record (as above) is developed in the data area vertex list describing the area represented by the vertex.

Relaxations on Square Finding and Results

In general, the program which finds square data areas by graph contraction performs its task exactly as discussed previously. However, in a subsequent route development step, too few paths or paths which had only excessive length may have been developed. At the same time, the implicit error in the locational precision of the data base may permit some relaxation in the interpretation of the accuracy of data boundaries and the associated impact assessments. This gives rise to the possibility of downward adjustment of some impact values during graph contraction. Accordingly, the five parameters described below may be used by the assessment team to relax the contraction process.

1. Relaxation is applied to vertices which have a different value from the current data vertices being contracted and which would cause the contraction to stop at that vertex. When the "expand" parameter is enabled, such a vertex value would be demoted, according to the conditions parameterized below, to the level of the area being contracted. This permits larger lower level impact areas to be developed.
2. The neighbor (gamma) parameter limits demotion to cases where the vertex which is a candidate for demotion may not have more than a specified number of immediate neighbors with the same value as it has. For example, if gamma was set to 0, the vertex must have no neighbors of its same value if it is to be demoted. This would permit only singleton areas to be demoted.
3. The value difference (delta) parameter limits demotions to cases where the value of the vertices being contracted and the value of the vertex which is a candidate for demotion differ by less than or equal a specified amount. For example, if delta was set to 1, then a vertex could only be demoted if its value differed by 1 from the current vertices being contracted.
4. The begin point (accept) parameter permits the two kinds of relaxations to be applied when the initial search is being carried out to locate a start vertex for graph contraction [as in Figure

10.4(b)]. For example, if the program was trying to locate an area with values of 2 to contract, it could be set to start with a 2 or a 3 which satisfied the demotion criteria.

5. The relaxation limit (level) parameter limits relaxations to be applied only to vertices with data values less than the specified level value. For example, if level was set to 5, only vertices with values from 1 to 4 could be considered candidates for demotion.

The relaxation which takes place is the result of the logical "AND" of the four conditions described. Examples of the effects of different combinations of relaxation parameters when applied to an arbitrary data set may be seen in Figure 10.5. A sample data set for a larger (30 x 30)

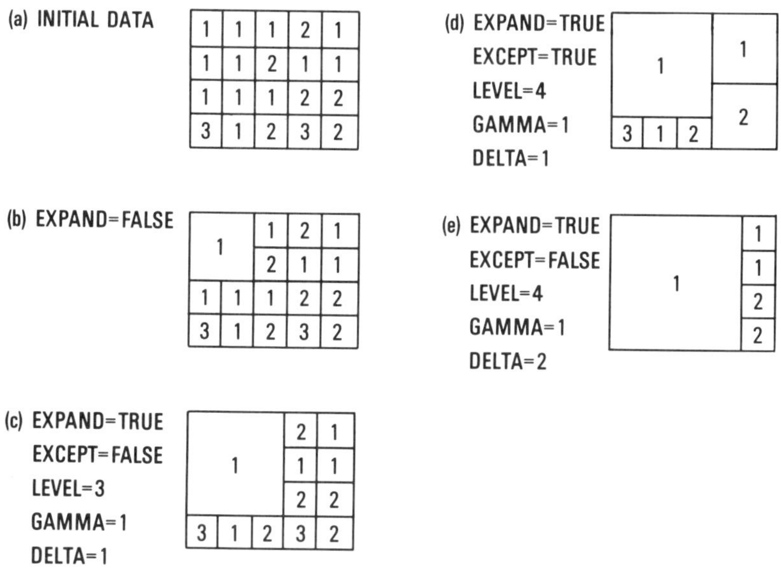

Figure 10.5 Relaxing graph contraction parameters.

area is shown in Figure 10.6. The data set was processed by the computer program using various parameter combinations. Figure 10.7 shows the results of unrelaxed graph contraction. Here the initial 900 areas (*i.e.*, vertices) have been reduced to 508; this represents a reduction to 56% of the original size of the data set while maintaining exactly the same data location coverage precision.

The effects of relaxations may be seen in Figures 10.8 and 10.9. In the figures demotion is permitted only if vertices differ in value by unity. In addition, Figure 10.8 permits each demoted area to have at most one

	1	2	3	4	5	6	7	8	9	10	11	12	13	14	15	16	17	18	19	20	21	22	23	24	25	26	27	28	29	30
30	5	5	2	2	2	1	3	3	4	6	1	1	1	1	1	2	3	3	3	3	6	6	6	5	4	5	5	5	5	5
29	5	5	2	2	2	2	3	3	4	4	1	1	1	1	1	3	3	3	3	6	6	6	6	5	5	5	5	5	5	5
28	4	3	2	2	2	2	3	3	4	4	1	1	1	1	1	3	3	3	3	6	6	6	6	6	5	5	5	5	5	5
27	4	4	2	2	2	2	4	4	5	5	1	1	1	1	1	3	3	3	3	6	5	6	6	5	5	4	5	5	5	5
26	1	1	2	1	1	1	3	3	4	5	1	1	1	1	2	4	4	5	5	6	6	6	5	5	5	5	4	5	4	5
25	1	1	1	1	1	1	3	2	5	5	1	6	6	6	4	4	4	5	5	5	6	5	6	5	5	5	5	4	5	5
24	1	2	2	2	3	3	2	2	5	5	3	6	6	5	4	4	4	6	6	6	6	5	6	5	5	4	5	5	4	5
23	1	2	1	2	3	3	2	3	4	4	3	3	3	4	4	4	4	6	6	5	5	5	5	6	6	5	4	4	4	4
22	1	1	2	2	2	3	2	3	4	4	2	2	2	3	4	5	5	5	6	6	5	5	6	5	6	3	4	4	4	4
21	1	2	3	3	2	3	3	2	1	2	2	1	1	3	4	4	5	6	5	6	6	6	5	1	2	3	3	4	4	4
20	2	2	3	3	2	3	1	1	1	3	3	2	1	3	4	5	6	6	5	6	6	6	5	1	2	3	3	3	4	4
19	2	3	3	3	3	2	3	4	4	4	4	4	2	5	5	5	6	1	1	1	6	6	6	4	4	4	4	3	4	4
18	2	2	2	2	2	1	2	3	4	4	4	4	4	5	5	6	6	6	6	6	6	6	6	4	4	5	5	5	5	4
17	3	3	3	3	3	2	1	2	3	4	4	4	4	4	6	5	5	5	5	6	6	6	6	6	6	6	6	5	5	5
16	2	2	3	3	2	1	2	3	3	3	3	4	4	4	5	5	5	5	5	5	6	6	6	6	6	6	5	5	5	2
15	1	1	2	2	1	2	2	2	2	2	4	4	4	5	5	5	5	5	5	5	5	6	6	6	6	6	5	5	2	2
14	2	2	1	2	1	2	2	2	2	2	2	4	4	4	5	5	5	5	5	5	6	6	6	6	6	5	5	5	2	2
13	1	2	1	1	1	2	2	2	2	2	2	4	4	4	4	5	5	5	5	5	5	6	6	6	5	2	2	2	2	2
12	1	1	3	3	3	2	2	2	2	4	4	4	4	4	4	4	5	4	4	4	5	6	6	5	5	2	1	1	1	3
11	1	1	3	3	3	3	3	1	1	1	2	4	4	4	4	4	3	3	3	4	6	5	5	5	2	1	1	1	1	3
10	2	2	3	3	3	3	1	1	1	1	1	2	4	4	4	4	3	3	3	4	5	5	5	1	1	1	1	3	3	3
9	2	2	2	4	1	1	1	1	1	1	1	2	2	4	4	3	3	3	3	3	1	1	1	2	2	2	1	2	3	3
8	3	3	4	4	1	1	1	1	1	1	1	1	2	2	3	3	3	3	3	3	1	1	2	3	3	2	2	2	2	2
7	3	3	3	4	1	1	1	1	1	1	1	1	2	2	2	3	3	3	3	2	1	1	2	3	3	2	2	2	3	3
6	3	3	5	4	6	1	1	1	1	1	1	1	1	2	2	2	3	3	1	2	1	1	2	3	3	3	3	3	3	3
5	2	2	5	5	6	6	1	1	1	1	1	1	1	2	2	2	3	3	3	1	1	1	1	1	1	1	1	2	3	3
4	2	2	5	5	6	6	6	4	4	4	1	3	3	3	2	2	2	2	2	1	1	1	1	1	1	1	2	2	2	3
3	2	2	3	3	3	6	4	4	4	4	4	3	3	3	3	2	2	2	2	2	1	1	1	1	1	1	2	3	3	3
2	1	1	3	3	3	4	4	4	4	4	4	3	3	3	3	2	2	2	2	2	1	1	2	2	2	2	3	3	3	3
1	1	1	2	2	2	4	4	4	4	4	5	2	3	3	3	3	2	2	2	1	1	2	2	2	2	3	3	3	3	3

Total Squares=900

Figure 10.6 Sample data set for routing examples.

neighbor of the same value. This resulted in some demotion and gave rise to a contraction to 439 areas in place of the original 900. This reflects a total reduction to 49% of the original size of the data set with some loss of precision. This represents a further decrease of 7% over the unrelaxed graph contraction.

Figure 10.9 shows the results of permitting vertices to be demoted which have two or fewer vertices of the same value. This resulted in additional demotion over the sample shown in Figure 10.8 and produced a contraction to 353 areas in place of the original 900. This is a total reduction to 39% of the original size of the data set with some increased loss in precision. This represents a further decrease of 10% over the example shown in Figure 10.8 and an increase of 17% over the unrelaxed graph contraction.

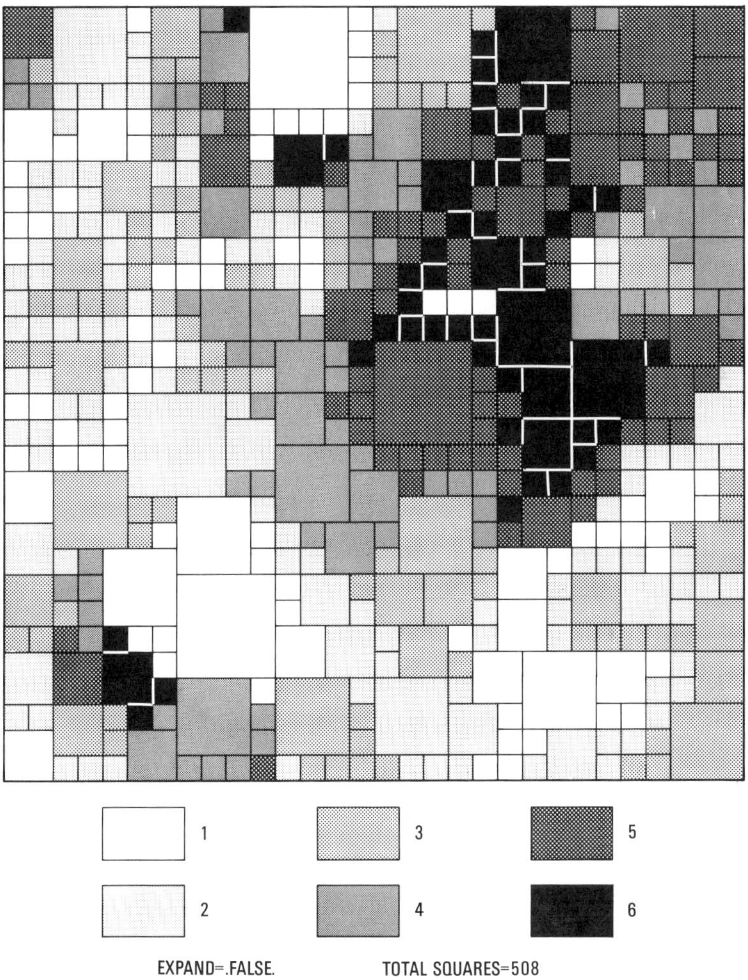

Figure 10.7 Unrelaxed square finding.

Contraction of the unit area graph causes a corresponding reduction in the size of its associated dual graph. This brings about a significant reduction in computing time required for route finding. This is due to the fact that the computing time required for the modified Dijkstra (1959) algorithm is dependent upon N squared when N is the number of nodes

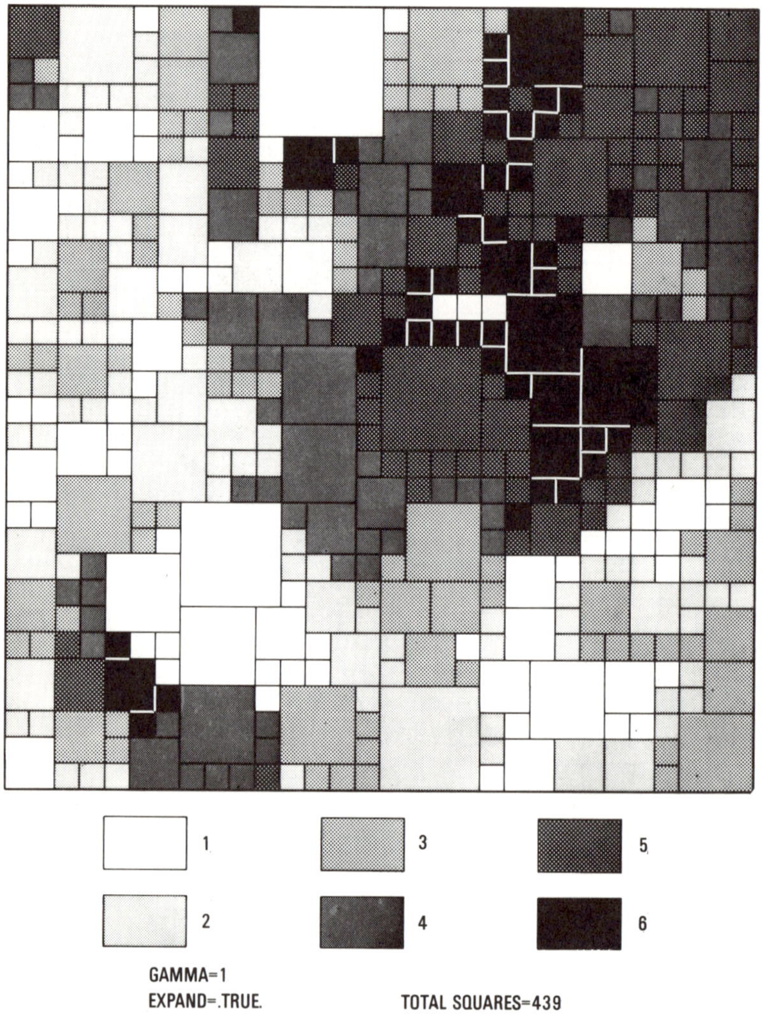

Figure 10.8 1 neighbor relaxation graph contraction.

in the dual graph (Pohl, 1969). Even a small reduction in the number of nodes brings about a significant reduction in computer time. For instance, in the previous example a reduction from 900 to 353 squares was achieved. This reduction to 39% of the original graph size brought about a reduction in computer time to 15% of the routing time in the 900 node

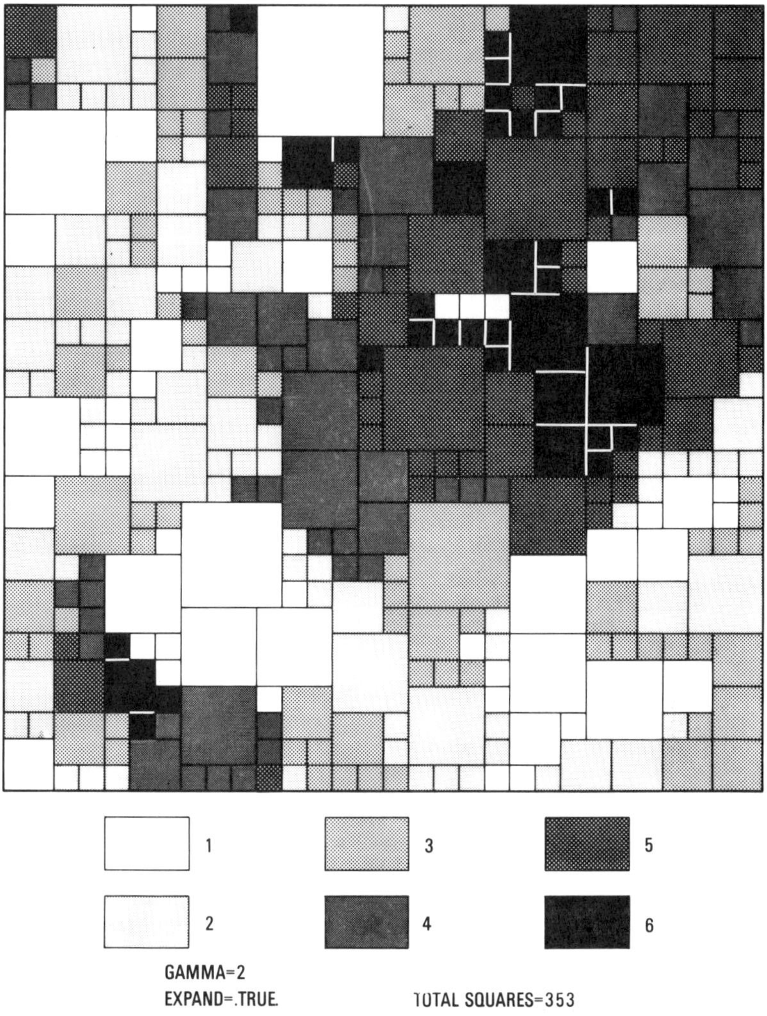

Figure 10.9 2 neighbor relaxation graph contraction.

graph. In a practical application to a graph of approximately 41,000 nodes, a reduction to 16,600 nodes was achieved. This resulted in a reduction to 16% of the computing time required for path finding in the noncontracted graph.

Interregular Area Distances and Adjacency

Essential to the development of routes through the dual graph is the appropriate intervertex distance. In the original unit area graph, these distances were related to the actual distances between the centers of adjacent areas. However, the "cost" of developing a route from one area to another is dependent upon the actual distance and the associated impacts over the distance. Potts (1975) reports that he and Goodchild replaced distance measures with the impact value directly in their assessment based on uniform square regions.

Graph contraction to regular (square) areas provides a range in square sizes over a study area. This, in turn, means that the cartesian distance from an arbitrary area to an immediate neighbor is not necessarily equal to its distance to another immediate neighbor. Figure 10.10 shows the

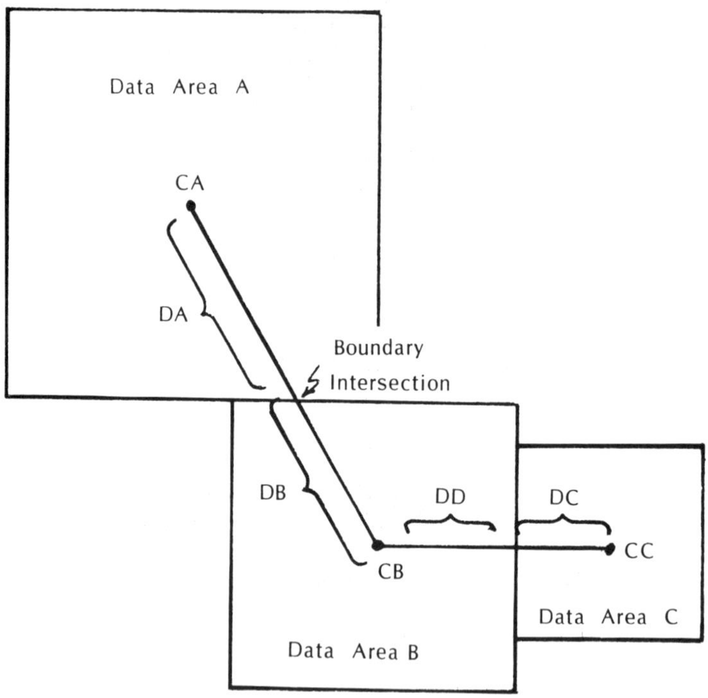

Figure 10.10 Interarea distance for nonuniform squares.

cartesian distance between the center of data area "A" denoted CA and the center of data area "B" denoted CB. It is seen to be made up of the two line segments with cartesian distance DA and DB. Similarly, the distance from CB to the center of area "C" denoted CC can be subdivided. It is clear from the figure that if data area A and data area C have the same impact ranking, the impact of route CB-CC is less than that of route CB-CA. This results from the distance DB being only slightly longer than DD while distance DA is much greater than DC. It is possible to represent the "cost" distance from CA to CB as:

$$\text{Cost} = \text{DA} \times \text{A impact} + \text{DB} \times \text{B impact}$$

Since "cost" is related only to the amount of areas A and B covered by the route and not the direction of the route, then the cost of route CA-CB is the same as route CB-CA.

An arbitrary mapping δ can be defined by the assessment team to map the "cost" of impact per unit distance from the impact rank of a unit area. Such a mapping can be described in tabular form as shown in Table 10.2.

Table 10.2 "Cost" per Unit Distance by Impact Rank

Impact Rank	Cost (a)	Cost (b)	Cost (c)
0	0	0	0
1	1	1	1
2	2	2	2
3	4	4	4
4	8	8	16
5	16	16	256
6	32	∞	∞

Three sample mappings are shown. The mapping shown in column (a) equates cost with impact rank as a power of two. This shows, for example, impact rank five as twice as costly as rank four. The mapping shown in column (b) is the same as (a) except "cost" for the prohibitive or no go rank 6 is declared to be infinite. The mapping shown in column (c) shows a very heavy cost being associated with high ranking impacts.

The "cost" for a connection between area i and area j denoted as C_{ij} can be expressed as a function of the mapping on the impact ranking for area i denoted r_i and the impact ranking for area j denoted r_j. Let b denote the point of intersection of the common boundary between areas i and j and the line i,j. Let D_{ib} represent the distance from the center of area i to the intersection point. Thus the function

$$C_{ij} = \delta\ (r_i)\ D_{ib} + \delta\ (r_j)\ D_{jb}$$

can be used to calculate the interarea distances.

Either there is a direct connection between a pair of unit areas or not. Where such a direct connection exists, the function C_{ij} is used to calculate the appropriate edge weight for the dual graph connection. It is equivalent in path development to consider an absence of connection between areas i and j as a connection between i and j with cost of infinity. This feature can be used when costing prohibitive impact (*i.e.*, no go) areas. Setting the cost map to infinity as shown in columns (b) and (c) of Table 10.2, would have the effect of completely removing the connections between any area coded 6 and any other area.

Where minimum corridor width is an important parameter in a study, extra criteria could be established when calculating C_{ij} values. For example, the length of the common boundary between adjacent areas could be examined to see if it exceeded minimum corridor requirements. If it did not, the value for C_{ij} would be set to infinity to indicate that no effective connection exists between area i and area j.

Development of Weighted Adjacency Matrix

This task is performed by a computer program which develops a matrix of C_{ij} values based on the "square" vertex list provided by the square finder procedure and the arbitrarily defined cost mapping for impacts. Previously it was noted that cost distances between two arbitrary areas are symmetric; that is, $C_{ij} = C_{ji}$ for all i,j. This means that only an upper triangular matrix need be developed to record all C_{ij} values. If it is assumed that C_{ij} values not defined are implicitly equal to infinity, then it is only necessary to store values where C_{ij} is not equal to infinity. This permits the use of sparse matrix techniques to store the weighted adjacency matrix. The list structure shown in Figure 10.11 is used by the adjacency program to store the matrix. The arrows show list address connections and ϕ indicates the end of the current sublist. Storage requirements are limited to the upper triangle by storing and accessing C_{ji} whenever C_{ij} is required and $i > j$.

ROUTE DEVELOPMENT

Route development is performed by a program based on a modified Dijkstra algorithm (1959) using as its input the adjacency matrix developed for the contracted dual graph. The source and destination locations for the utility route are arbitrarily defined by the assessment team. These terminal points may have been specified by the proponent or may be a

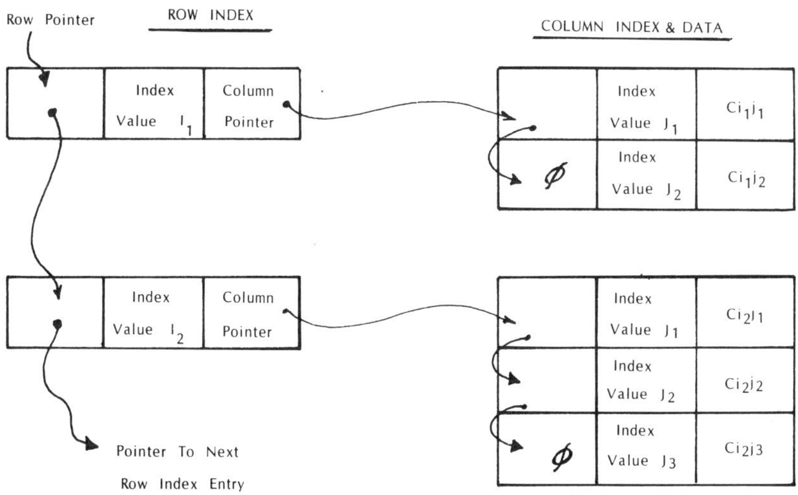

Figure 10.11 Sparse matrix storage for adjacencies.

set of trial points being evaluated by the assessment team. If more than one pair of terminal points is to be studied, the routing analysis is re-applied for each required combination of terminal points.

Dijkstra Algorithm Operations

The basic algorithm used by Dijkstra is outlined in Chapter 2. The modifications added develop all alternate routings of equal length while the algorithm proceeds. Figure 10.12 shows a simple path routing problem. Vertices in an arbitrary graph are designated by letters from A to I.

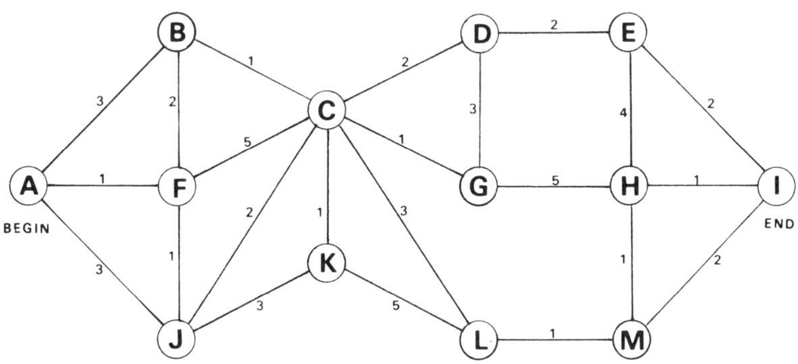

Figure 10.12 A sample routing problem.

The edges connecting the vertices are not directed and a weight or cost is specified for each. The Dijkstra algorithm solves the problem: given a source vertex (say, A) and a destination vertex (say, I) find a minimum cost or weight path between them. Essentially the algorithm proceeds by developing a solution set S. To begin with, the vertex A is added to the solution set. The algorithm iterates extending the membership of set S by adding vertices. At each iteration, the coboundary of the subgraph generated by set S is examined to locate an edge in the coboundary which provides the least-cost extension to a least-cost path which terminates with a vertex in S. The algorithm terminates when the destination vertex is moved into S or there are no vertices left which can be added to S. In the latter case, no path exists.

As Dijkstra's algorithm proceeds, a route graph can be established based on the step at which a vertex was added to S. Figure 10.13 shows the route graph developed by applying Dijkstra's algorithm to the example graph in Figure 10.12. Vertices are specified by their identifying letter and the number of the iteration when they were added to S. An arrow from vertex to vertex indicates a route segment was developed in that direction. The number pair on each edge identifies the actual edge cost or weight in the numerator position and the total cost or weight associated with reaching the destination vertex for this edge from the source vertex in the denominator position. Since the algorithm works from edges in the coboundary of S, edges which connect a vertex in S to another vertex in S are ignored.

Figure 10.13 shows that destination vertex I is reached at step 13 with a total cost of 10. The route followed is the vertex sequence:

A, F, J, C, D, E, I

However, it is evident on inspection that at least three more paths of the same length exist which connect the source to destination. In the environmental routing problem we wish to identify all "reasonable" alternate routings. This should include at least all minimal equal cost routes from source to destination. The algorithm has been modified accordingly.

Modifications to Dijkstra's Algorithm

In the revised algorithm, a set F is used in a similar way to develop routes. A new set R is introduced such that

$$\{S = R \cup F\}$$

when S is the solution vertex set of Dijkstra's algorithm. Initially, R = ϕ and F = A when A is the source vertex. New edges are added to F

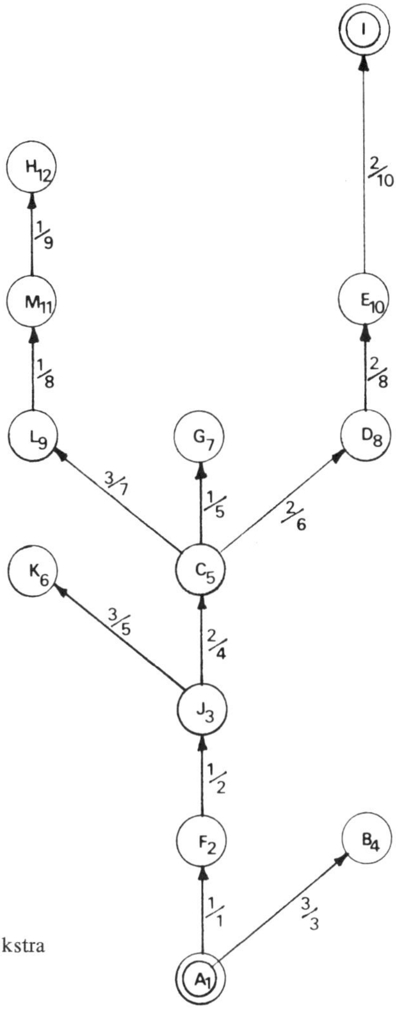

Figure 10.13 A Dijkstra route graph.

as per Dijkstra's method. However, the revised algorithm proceeds to inspect all edges incident to members of S which would provide the least cost route extension to the least cost route in R ∪ F. Since this set can include edges which are incident to two members of F, potentially the number of edges for consideration are increased slightly over Dijkstra's algorithm. However, by maintaining R, F and D (the distance–adjacency matrix) in suitable list structures where elements are located in their respective list by their distance or accumulated distance, algorithm speed is improved. By using this list strategy, the algorithm can locate at each iteration, the appropriate vertex to extend the route from and the edge to use without a search. This modified algorithm has exhibited a time requirement on the order of the square of N in tests. When such an edge is selected, it is added to the route graph only if the cost to reach the edge destination vertex is equal to the already established least-cost route to that same vertex. Once added to F, a vertex remains in F until all edges incident to it are considered and added to the route graph or rejected. When no edges are left for consideration, the edge is removed from F and placed in R. F corresponds to the active frontier of route development and R contains the vertices which define the already established and nonexpandable subroutes.

Figure 10.14 shows the results of applying the modified algorithm to the sample graph in Figure 10.12. As described previously, vertices are identified by letter and the iteration step where they are added to the route graph. In this case, however, some vertices have two numbers. This indicates that the iteration step number where each of the two edges were added to the graph. For example, vertex B was added to the graph at step 4 by the edge running from A to B, and the edge from F to B was added at step 6. The algorithm iterates until no edge can be selected incident to a vertex in the frontier F which yields a route with cost equal to an existing route to the destination vertex. The example shows that the following routes exist connecting vertex A to vertex I with path cost 10:

(a) A, F, J, C, D, E, I
(b) A, F, J, C, L, M, I
(c) A, F, J, C, L, M, H, I
(d) A, B, C, D, E, I
(e) A, B, C, L, M, I
(f) A, B, C, L, M, H, I
(g) A, F, B, C, D, E, I
(h) A, F, B, C, D, E, I
(i) A, F,B, C, D, E, I

In the figure, some vertices are shown with partial edges terminated with crosses and with vertex letter indicated. This indicates an edge connection was considered between two vertices when both were members of F and the edge was rejected because it would provide a path length greater than the existing path to the destination vertex for the edge. For example, an edge was not extended from vertex D to vertex G because the path cost to G through D would have been 9 while G was already reached by a path from C with a cost of 5.

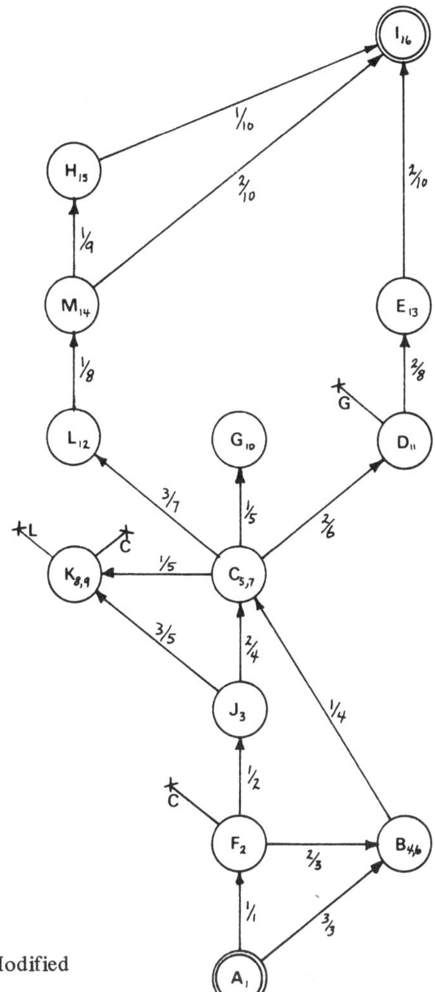

Figure 10.14 Modified route graph.

Extension to Modified Algorithm

Two extensions have been added to facilitate alternate route development in the planning system. The program which implements the modified Dijkstra algorithm can be reapplied by the assessment team specifying, by means of a parameter, that N extra routes are to be located. This is achieved by simply forcing the program to continue execution beyond its normal termination point. Of greater importance is the extension which permits the algorithm to accept as equal length subroutes, routes whose aggregate costs differ by less than an arbitrary epsilon value. This epsilon value is essential to reduce any computational round-off problems which result from the calculation, in real arithmetic, or interarea connection costs. In addition, larger epsilon values can be used to permit the development of alternate routes when it is known that the base impact assessments are not precisely accurate.

ROUTE STRAIGHTENING

As a result of applying the routing algorithm discussed in the previous section, the study team obtains lists of vertices in the dual graph which represent sets of least-cost routes. If the dual of the dual graph is taken at this point, the sequence of route vertices can be considered as the connection between the centers of the corresponding areas. The resulting route can be quite irregular as shown in part (b) of Figure 10.15. However, the unit area for impact assessment was selected as the smallest area unit required for locational precision and, accordingly, can be considered homogeneous in nature. This makes it appropriate to shorten path length by rerouting the path through the areas independent of square centers. This is achieved by the route-straightening algorithm.

Route-Straightening Algorithm

The algorithm is applied to the contracted unit area graph based on the least-cost path identification provided by the route development program. The algorithm may be described in general terms as follows.

A. The straightened path is developed from the center of the source unit area to the center of the destination unit area.

B. Assume that point p_i lies on the straightened route and p_i is contained in some contracted area A_i: Determine the successor point p_{i+1} which also lies on the straightened route by applying the following steps:

1. Determine the points R,S, on the common side segment between area A_i and A_{i+1} then A_{i+1} is the successor area to A_i as determined by the original irregular least-cost path.

2. Project from p_i through R,S two lines until they meet at T,U the furthest boundaries (*i.e.*, sides) of area A_{i+1}. [See part (a) of Figure 10.16.]

3. If part of the farthest sides of area A_{i+1} contained between points T and U are common to an area A_{i+2} which is the successor area to A_{i+1} as determined by the original irregular least-cost path, then redefine R and S to be the end points of the common side between T and U and area A_{i+2}. Now reapply steps 2 and 3 using

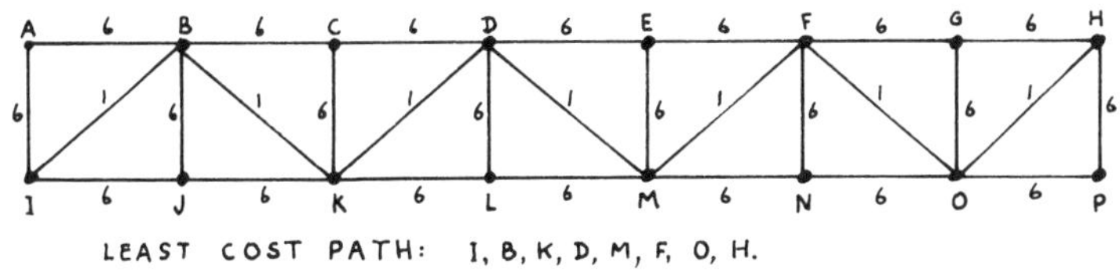

(a) Dual graph

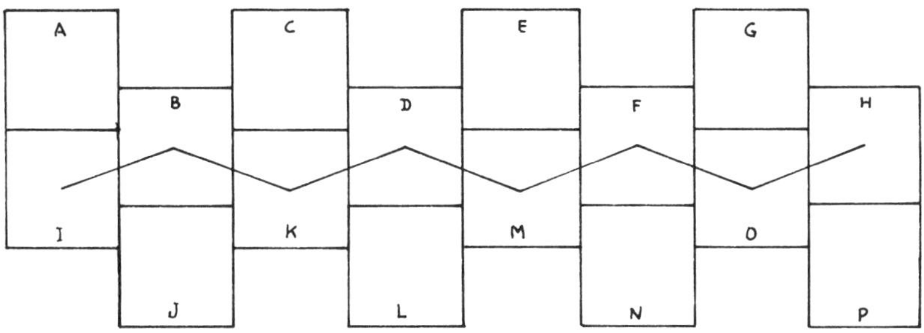

(b) Associated unit area graph

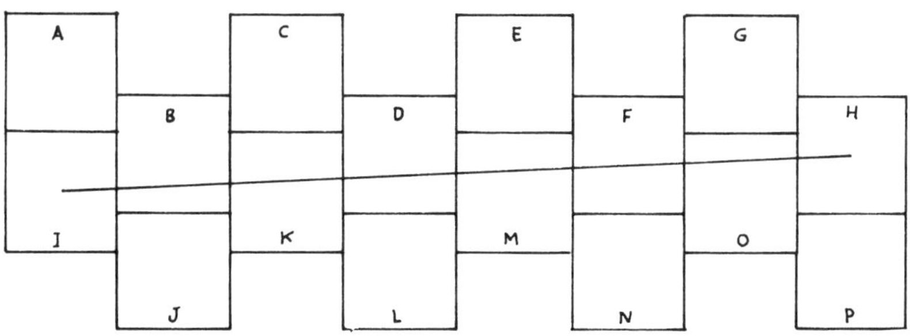

(c) Straightened route

Figure 10.15 A degenerate irregular route.

the area A_{i+j} when j is the number of times these steps are repeated. [See part (b) of Figure 10.16.] Otherwise proceed with step 4.

4. The iteration of steps 2 and 3 has established A_{i+j}, $j \geqslant 1$, and area containing the farthest projection of the solution area from p_i. Now initially with k = 1 consider the center of area A_{i+j+k}. Apply the same procedure (calling R and S, E and F, respectively) as step 1 and 2 in the reverse direction using the center of area A_{i+j+k} as point p_i. [See part (a) of Figure 10.17.]

5. Determine the intersection of the final solution space derived at step 3 with the backward solution space obtained in step 4. If the

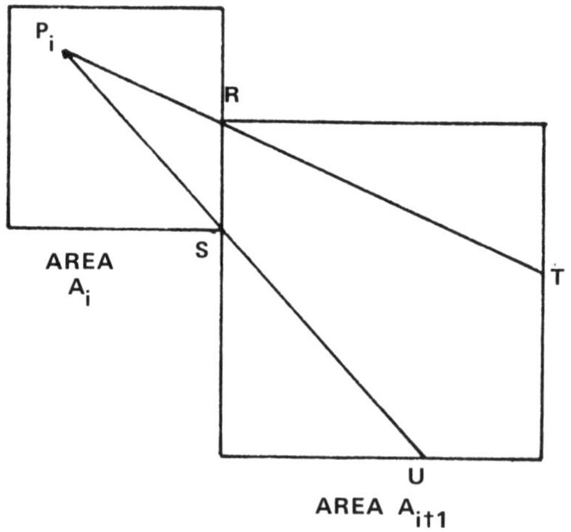

(a) Projection of solution area via common line segment

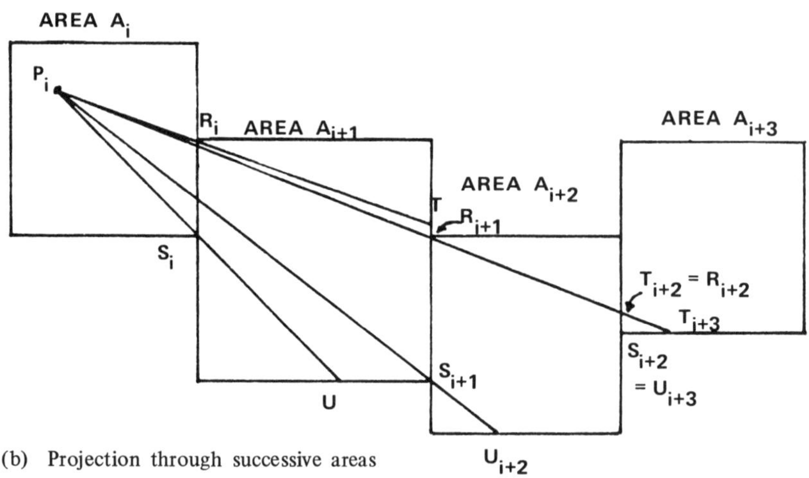

(b) Projection through successive areas

Figure 10.16 Forward projection of solution area.

intersection is not empty, set k = k + 1 and reapply step 4. If the intersection is empty, go to step 6.

6. Redefine k = k - 1. Define as point p_{i+1} the R or S value farthest away from p_i such that p_{i+1} is closest to C_{i+j+k}.

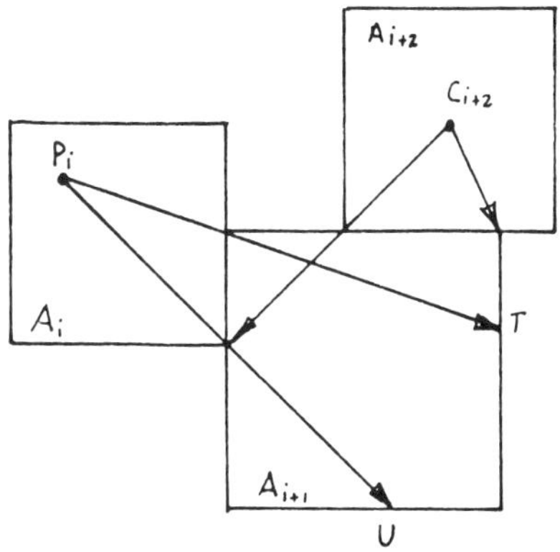

(a) First stage backward projection from C_{i+2}

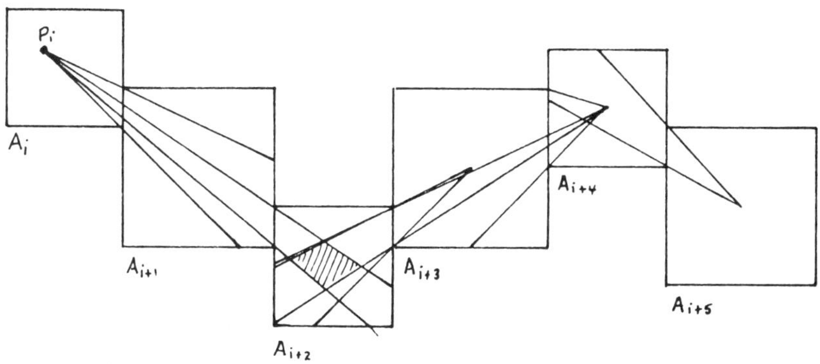

(b) Two area forward and two area backward projection

Figure 10.17 Backward projection.

7. Reapply steps 1 through 6 using as p_i the new point p_{i+1} until p_{i+1} is the center of the destination area.

Part (b) of Figure 10.17 shows the solution area definition which led to the determination of new point p_{i+1} from points p_i and C_{i+4}. Figure 10.18 shows that the algorithm successfully straightens the degenerate case given as an example in Figure 10.15.

CORRIDOR ASSESSMENT

The routing algorithms specify a series of points in space which, if connected, provide minimum-cost routes from an arbitrary source to an

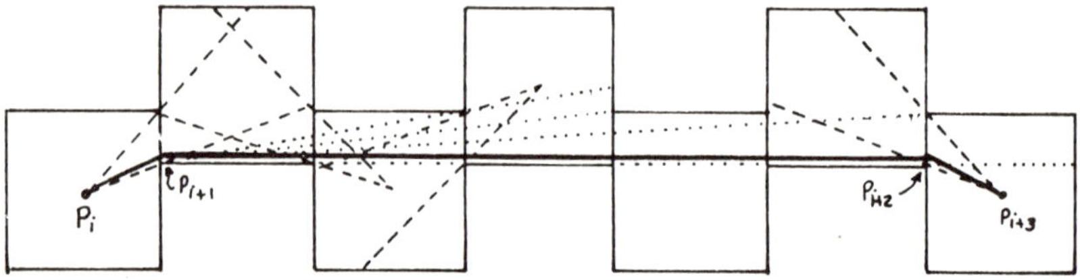

Figure 10.18 Straightening a degenerate route.

arbitrary destination. While each such route may have an aggregate impact cost which is equal to all other routes, it is quite possible that routes may be quite dissimilar one from another with regard to the kinds of impact each possesses. A corridor assessment procedure is included as the last analysis step in the planning system to permit intercorridor impact comparisons. Thus, not only alternative corridors are specified but their characteristics are tabulated to facilitate external review.

Corridor assessment involves examining the spatial extent of corridors with respect to the data base of impact factor rankings and raw data. The planning system's corridor assessment program uses as input: the series of rank location points provided by the routing program, the unit area data base, the unit area impact rankings by factor and composite, and set of weighting (or costing) parameters arbitrarily set by the impact assessment team. The program provides, for each corridor identified in the routing phase, a tabulated output listing the actual area (usually in acres) of each specified parameter involved in the corridor. In addition, the program generates a plotter file which can be used to overlay the spatial extent of each corridor or any of the data, impact factor and/or composite maps produced in the study. This is provided to facilitate external review.

Corridor Assessment Process

The impact assessment team specifies the corridor width required for the utility under consideration. The program transforms the set of route center points obtained from the routing study to a set of corridor polygons. Part (a) of Figure 10.19 shows a sample set of route center points as they relate to study unit areas. One half the specified corridor width is assigned each side of the straight line connecting route points. This yields a set of parallel lines which serve as the sides of corridor polygons as shown in part (b) of Figure 10.19. Each straight line of part (a) is bounded by parallel polygon boundaries which are the specified corridor

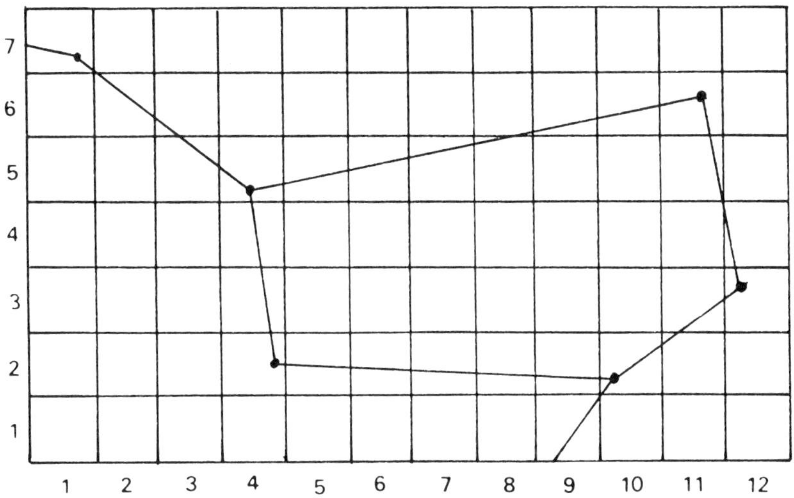

(a) Unit areas and route center points

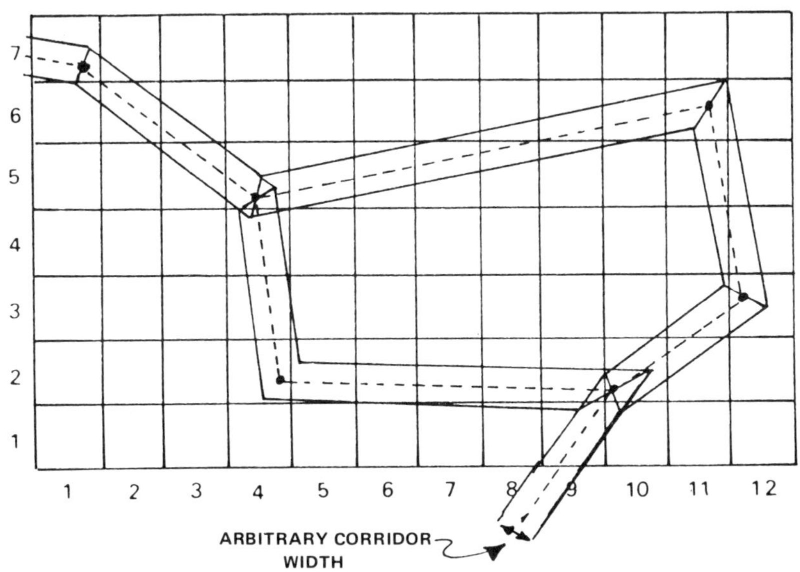

(b) Unit areas and corridor polygons

Figure 10.19 Corridor assessment.

width wide. Due to the variability in angle between successive route segments, the corridor polygons have nonparallel ends where corridor segments join; all corridor polygons are four-sided.

Once the corridor polygons are defined, each unit area can be examined to determine the percentage of its area occupied by the corridor. The program examines separately each corridor, determining each unit area and the associated areas which are included in the corridor. This is achieved by calculating the spatial intersection between the corridor polygons and the polygons representing the unit area graph. All corridors are examined before any assessment totals are calculated. Each unit area involved in a corridor has a record associated with it placed in a disk file. The record consists of: (a) X coordinate of the unit area; (b) Y coordinate of the unit area; (c) corridor identification; and (d) percentage of the unit area in the corridor specified. If a unit area is traversed by more than one corridor [for example, see unit areas (4,5) and (10,2) in part (b) of Figure 10.19], an entry is made for that unit area in the file for each corridor.

When the examination of each corridor is complete, the disk file of affected unit areas is sorted in the same order as the data bank by coordinate to force all corridor records for each unit area to be clustered together. The sorted file is used by the program as a set of retrieval keys to access the study data base. Since all unit record access requirements are clustered in the sorted file, all corridors are assessed in one pass of the data base.

The assessment team indicates which items or combinations of items in the study data base are to be tabulated, and it may apply a set of weights (or costings) to scale the tabulations. The program develops, concurrently, a separate tabulation for each corridor being assessed. Figure 10.20 shows such a sample tabulation. This tabular output is repeated successively for each corridor being assessed. The actual tabulations can be presented during external review along with the plotted outline of the corridors on appropriate study maps.

An Example Assessment

A sample impact data set is presented in Figure 10.6. Three contractions of this sample data set which result from computer program graph contraction are shown in Figures 10.7, 10.8 and 10.9. When the arbitrary source points 15 and 49 and destination points 305 and 216 are specified, Figure 10.21 shows the basic routing established. Figure 10.22 shows the straightened routes and Figure 10.20 shows part of the summary for one of the corridors in Figure 10.22.

ANALYSIS OF PATH

PATH ID IS TEST RUN

CLASSIFICATION	PERCENTAGE CORRIDOR	AREA ACRES	PER/UNIT VALUE	COST
FIRST LAND USE				
BUILT UP AREA	.1246E+00	.4984E+00	4.00	.08
OUTDOOR RECREATION	.2126E-01	.8503E-01	4.00	.00
WOODLOT	.5414E-01	.2166E+00	4.00	.00
SECOND LAND USE				
BUILT UP AREA	.1247E+00	.4968E+00	4.00	.03
OUTDOOR RECREATION	.1433E+00	.5734E+00	4.00	.26
WOODLOT	.3195E-01	.1278E+00	4.00	.01
THIRD LAND USE				
BUILT UP AREA	.7259E-01	.2903E+00	4.00	.10
OUTDOOR RECREATION	.1884E+00	.7536E+00	4.00	.24
WOODLOT	.2390E+00	.9561E+00	4.00	1.97
FIRST ZONING USE				
INSTITUTIONAL - HOSPITALS, SCHOOLS, + NURSING HOMES FOR THE AGED	.1218E+00	.4873E+00	4.87	.29
ORBITAL LAUNCHING SITE	.3282E+00	.1313E+01	4.00	18.38
SECOND ZONING USE				
ESTATE RESIDENTIAL	.9419E-01	.3768E+00	4.00	.18
INSTITUTIONAL - HOSPITALS, SCHOOLS, + NURSING HOMES FOR THE AGED	.1558E+00	.6232E+00	23.99	2.33
THIRD ZONING USE				
ESTATE RESIDENTIAL	.1870E+00	.7479E+00	4.00	4.71
INSTITUTIONAL - HOSPITALS, SCHOOLS, + NURSING HOMES FOR THE AGED	.1130E+00	.4521E+00	8.14	.42
FIRST NATURAL RESOURSE				
FISHERY	.3116E+00	.1246E+01	4.00	30.98
GRAVEL PITS + QUARRIES	.1884E+00	.7536E+00	4.00	3.00
SECOND NATURAL RESOURSE				
FISHERY	.4509E-01	.1804E+00	4.00	.14
CAMPGROUNDS PUBLIC + PRIVATE	.1507E+00	.6028E+00	4.00	1.86
GRAVEL PITS + QUARRIES	.2042E+00	.8168E+00	4.00	9.84
THIRD NATURAL RESOURSE				
FISHERY	.1127E-01	.4509E-01	4.00	.00
CAMPGROUNDS PUBLIC + PRIVATE	.5105E-01	.2042E+00	4.00	.17
GRAVEL PITS + QUARRIES	.3768E-01	.1507E+00	4.00	.03
TOTAL	.9734E+00	.1200E+02		75.01

TOTAL AREA OF PATH IS .393122027993377E+01

Figure 10.20 Sample corridor tabulation.

Figure 10.21 Sample route selection.

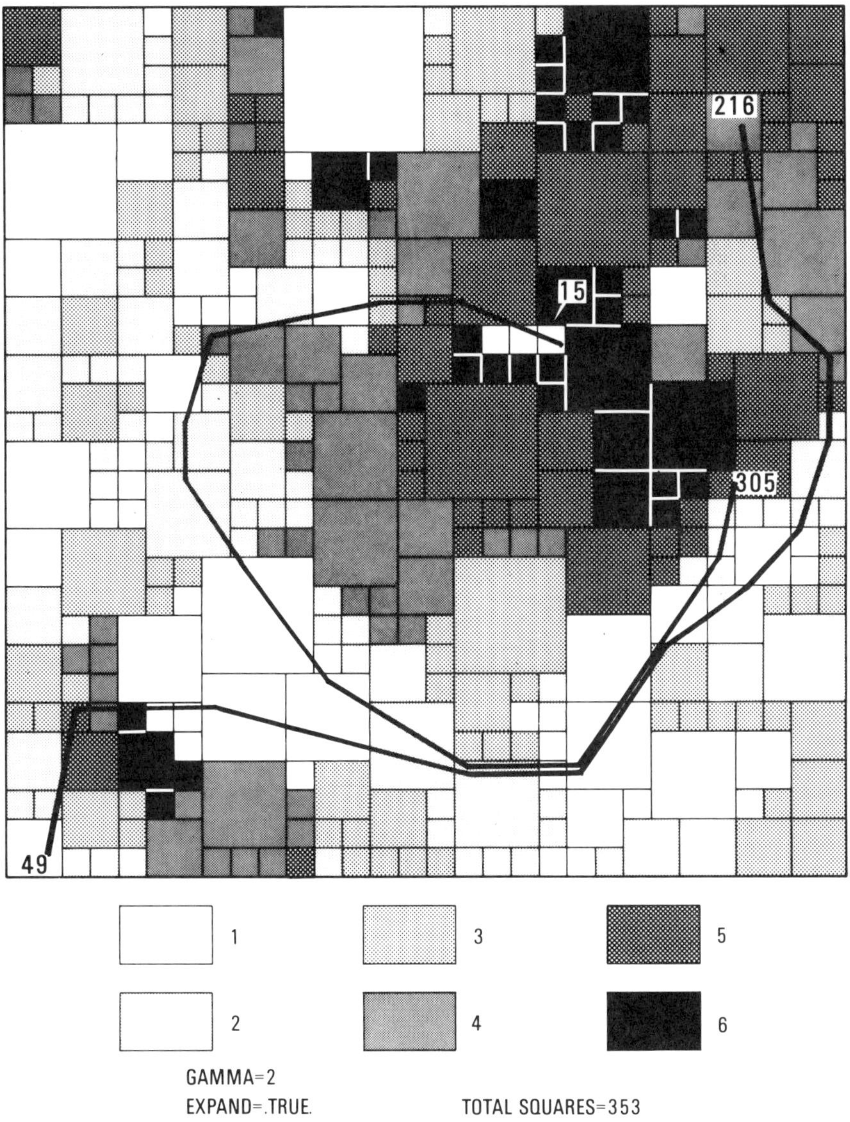

Figure 10.22 Sample route straightening.

CHAPTER 11

A SAMPLE APPLICATION

INTRODUCTION

In Ontario, the electricity utility company Ontario Hydro planned to construct a new bulk power transmission system. New thermal and nuclear generating stations were to be linked into a 500-kilovolt grid. The projected transmission network constituted a major addition to the provincial energy distribution system. One of the bulk transmission system segments had been planned to run between the Nanticoke generating station and London, Ontario. The Ministry of the Environment's Green Paper (1973) on government policy supports a general requirement that planning for each part of the new network had to be subjected to environmental impact assessment and full public examination. Impacts on social, economic, cultural, ecological and visual-aesthetic factors were to be assessed.

Ontario Hydro retained the consulting firm of James F. MacLaren Ltd. to undertake such a study. The basic terms of reference specified by Ontario Hydro may be paraphrased:

> "The objective of the study is to develop alternate power corridors between the end points and to refine to a specific route, the one which, in general, follows the path of minimum impact on visual, socio-cultural, economic and ecological requirements; this route having been determined with the benefit of public input throughout the study area."

The following constraints and assumptions were identified by Ontario Hydro:

1. The alternative corridors must be broadly compatible with the existing and overall proposed systems of Ontario Hydro. The chosen corridor is to be one element of a grid system of wider extent, and the ultimate choice among alternative routes may be governed by this external consideration.

2. The visual impact is a function of the dimension of the hydro towers and transformer station(s). The towers could be at least 162 feet high and with strung cables are highly visible objects. The lines will be in series of two or three abreast requiring a right-of-way between 425 feet and 600 feet in width. Visibility may, however, be reduced by configuration of the land, and by the relative position and attention to natural vegetation.
3. Socio-cultural impact is partly a function of visibility with the tower line having potential impact on aesthetic and scenic attributes of the landscape. Direct impact on specific sites is anticipated, *i.e.*, residential and historical localities. An impact on property values may also be anticipated. It is noted, however, that the corridors cannot include any major built-up residential areas. The impact will therefore be concentrated mostly in rural areas whose relative values will have to be determined.
4. Economic land impact will be in respect to other economic uses, *i.e.*, agricultural, commercial, industrial, etc. In some cases, including other utility and wave transmission systems, there will probably be complete incompatibility. In other cases a range of partial impacts may be expected.
5. The potential ecological impacts will be in terms of the swathe which any corridor would cut through existing ecosystems. The varying effects on different landscapes and habitats, and the potential dangers to migrating birds are factors to consider.

A study team in the firm of James F. MacLaren was established consisting of S. H. Janes, A. R. Glasgow, and J. E. Dooley. In addition, the firm made a research grant to R. T. Newkirk (as principal researcher) and M. J. Troughton at the University of Western Ontario–the terms of reference being to develop and implement the computer-based environmental study for the project.

Newkirk designed the general system and all of its computer algorithms. He directed the implementation of the computer algorithms and the day-to-day task of data digitizing and interpolation. Part of the system had been implemented by the author before the grant from James F. MacLaren was received. Notably, the interpolation procedure for data input and the programs to draw the data certification maps and impact factor maps were already developed. Other parts of the system were designed and implemented by the author during and after the period of the three grants associated with this application. Some were developed to meet requirements discovered during the application (notably the Cascade Algorithm); others were developed later to be part of the general planning system but not used in the application (the graph contraction and routing algorithms). The requirement for the Cascade Algorithm became evident when composite impact by weighted average was found unacceptable. J. E. Dooley proposed impact combination by maximum. This was implemented, but concern was raised by Ontario Hydro and others that threshold conditions needed to be accommodated. In response to this need, Newkirk formalized and implemented the Cascade Algorithm.

During the application, the members of the James F. MacLaren study team designed and implemented a survey of public opinion. They attended public meetings and met with officials of Ontario Hydro and various private and governmental agencies to determine major concerns to be included in impact assessments. They interacted with Newkirk and Troughton to: identify the specific impact factors to consider, select data base variables required for the application, and define the impact factor calculation "rules." In addition, they made the necessary arrangements to locate and obtain necessary data series not in the public domain. In some cases they assembled special manuscript map series. An important role of the MacLaren team was to carry the results of the study to Ontario Hydro, the government and the public. Newkirk joined the MacLaren team when the study was presented to the government's external review committee.

In addition to assisting in the selection of impact factors for this application, M. J. Troughton took major responsibility for locating sources of relevant data in the public domain. He was in particular responsible for specifying and refining the agricultural operations impact factor.

Based on its internal system planning and integrity studies, Ontario Hydro established the boundaries of the study area to be considered. In addition it specified some source and terminal points for the potential transmission line. The actual straight-line distance between source terminal points was slightly over 60 miles. Depending upon tower technology used it can be safely estimated that line construction costs could run between \$200,000 and \$700,000 per mile exclusive of system integration costs (*e.g.*, switching and transformer stations). Thus, line construction costs would, at a minimum, be \$15 to \$25 million.

The impact study was carried through the first three planning system phases up to the development of a composite impact assessment during 1973-1974. Late in 1974 the Ontario Government requested Ontario Hydro to suspend the development of its bulk (500-kV) power grid pending review and evaluation of its need by a Royal Commission chaired by Dr. Arthur Porter. Accordingly, the planning project for the Nanticoke to London transmission line was suspended awaiting government approval to proceed. The Commission had not completed its report by fall 1977 and the projects remained on "hold" status at that time. In this chapter, some salient features of the study are outlined. More description may be found in Janes *et al.* (1974), Newkirk and Dooley (1976), and Newkirk and Troughton (1975).

STUDY AREA AND ANALYSIS UNITS

The study area as specified by Ontario Hydro is an irregularly shaped area of 2700 square miles in extent as shown in Map 7. It includes all or part of 6 counties, 13 urban municipalities and 43 rural municipalities including 4 Indian Reserves. The total population of the area is approximately 430,000. Overall, the area is a subregion of the primary corridor of socioeconomic development within Canada. The study area, although fairly well documented, posed many problems consistent with its position in a densely populated, complex region. Some of the problems include:

a. a dense pattern of intensive land use, including urban industrial developments, intensive agriculture, power, energy and communication facilities;
b. complete, largely private, ownership of land and a finely divided mosaic of different types of land use and different sizes of land parcels;
c. pressure on rural land area to supply not only a continuing base for any specialized, capital-intensive, types of agriculture, but also to service active and passive recreation needs and to provide an attractive "countryside" as the hinterland for growing urban centers;
d. a generally flat, "subtle" landscape whose low variability of relief leads to potentially high visibility of the 500-kV line, and in which the relatively few areas of variable topography have high value from aesthetic, scenic, ecological and recreational points of view;
e. although possessing no insurmountable barriers to transmission line construction, local factors of steep slopes, river crossings and wetlands posed possible constraints to siting towers and transformer stations.

An evaluation of data sources revealed that the most useful data were at a map scale of 1:50,000. Accordingly, a basic analysis unit of 500 meters was established. Fortunately, a collection of 14 physical landform, soil, elevation, and drainage variables had been recorded for a significant part of the study area by McDaniel *et al.* (1970). These data were added to the study data base after extensive updating and certification. During the study, 15 complete map series consisting of 13 single map sheets each were assembled for the study area and input to the data bank using the map sheet digitizing and interpolation process.

The computational data base, organized on a 500-meter grid square basis was keyed to the basic 1,000-meter grid on the 1:50,000 topographic map series. This resulted in the study area being divided into approximately 35,000 unit areas of roughly 60 acres each. For each such 500-meter grid square, an extensive record was developed including information on physical characteristics, land use details, land use zoning, use by utilities, natural resource inventory, capabilities to support recreation wildlife, etc. The initial data base was developed over a six-month period

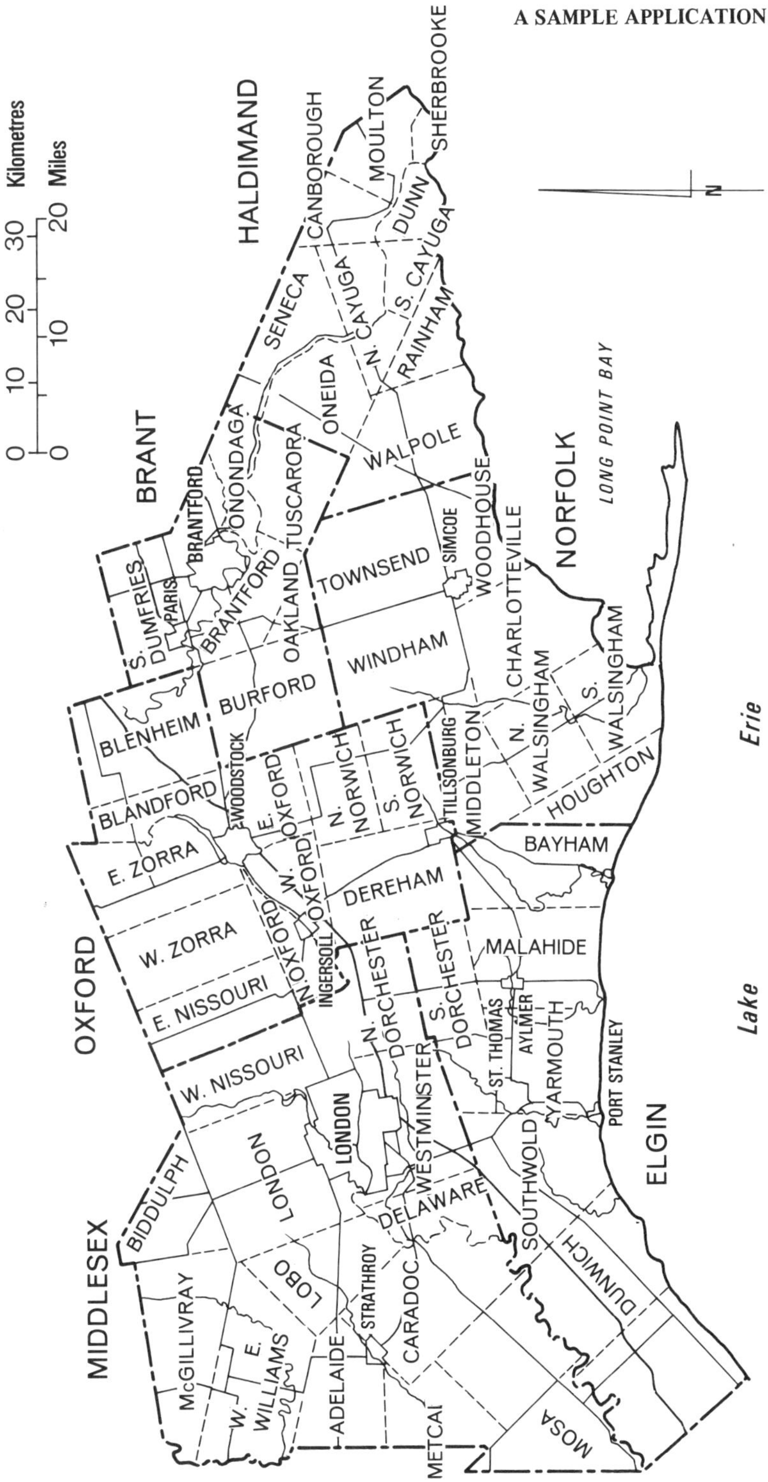

Map 7. Study area.

and consisted of 77 information items for each 500-meter unit area. In order to provide even finer analysis capability related to land use, up to three multiple uses within any 500-meter region were recorded as percentages of the basic analysis unit. Land uses covering as little as four to six acres were recorded.

IMPACT FACTOR DEVELOPMENT

Central to the research method is the development of impact factors; in this application eight impact factors were derived in steps. Initially, expert opinions of researchers at the University of Western Ontario, staff of James F. MacLaren Ltd. and Ontario Hydro were combined with public opinion derived from questionnaires and public meetings. An initial set of factors and a data base were developed and impact maps produced. Representatives of various ministries of the Ontario Government reviewed the results and made many sound and constructive criticisms. A final set of factor definitions was developed by the study team with ministry advice and additional data.

In particular, James Dooley and A. R. Glasgow developed the initial definition of factors four, five and six. Dooley developed the initial revision of all factor rules based on governmental review. M. J. Troughton and R. T. Newkirk developed the initial definition of factors one, two, three and seven. Newkirk defined the initial specification of impact factor eight. Newkirk revised all initial impact factor "rules" to refine the assessments. The implementation of the impact factor algorithms was directed by Newkirk.

Data base variables required to evaluate the initial impact factors were assembled, the impact maps drawn and subjected to governmental review. In that review, government agencies saw the requirement for more detailed information than that available to the study team and made the appropriate confidential data series available. This gave rise to a revision to the original factor rules. One impact factor was completely replaced. After the review of the first map series, the restructuring of factor rules, determination and evaluation of new data sources, addition of new data to the data bank, impact factor calculation and map production were completed in a little under three months. The system's flexibility is evident.

The final list of data variables was extensive. Many of the variables involved a number of possible classification codes. Wherever possible, the full classification provided on the source document was input to the data bank. Appendix I lists the final variables and codes which made up the data base for the study.

The final impact factors are thought to reflect public and governmental impact concerns, utilizing suitable data. The final set of eight impact factors are:

1. Rural Operational–an assessment of the surface of potential impact on agriculture and associated rural activity. The factor combines measures of soil capability for agriculture and actual land use, future land use, dollar values of production and assessment of investment in tobacco operations. This factor represents a combination of present and future agricultural value of the land.
2. Natural Environment–an assessment of the impact that a transmission right-of-way would have on the communities of flora and fauna in the natural environment. Data analyzed to evaluate this factor included: location of biologically sensitive areas as defined by the Ministry of Natural Resources, woodlot cover, and terrestrial, aquatic and wetland wildlife suitability as classified by the Ministry of Natural Resources.
3. Recreational, Cultural and Historical–an appraisal of the region in terms of the combination of actual and potential recreation and the inclusion of areas designated cultural and historic by federal, provincial or local authorities.
4. Residential, Commercial and Institutional–compiled record of these land uses, mostly representing areas ruled out for use by a hydro corridor were assembled into a single factor. Built-up urban areas, nonurban built-up areas and planned residential land use were included in the factor.
5. Linear Utilities–identifies and rates systems of other transportation and utility networks which cross the area and identifies where restrictions on paralleling or problems with crossing may occur. On the basis of information supplied by highway, railway, oil, gas and water companies or agencies, the data were mapped and included in the data base for this factor.
6. Area Utilities and Industry–locations of area utilities such as airports, wave transmission and reception facilities and their zones of influence were mapped along with industrial sites to provide a data base for this factor.
7. Natural Landscape Diversity–an assessment of the relative value of nonflora or fauna natural landscape through a diversity count. Twelve different features were investigated in each 500-meter cell for presence or absence. A high count indicated high value area and low count the opposite.
8. Relative Visibility–a topological measure based on elevation relative to the surrounding area and adjusted by vegetation cover. The assessment provides for each 500-meter grid square, a measure which indicates how visible a tower placed in that square would be when viewed from the surrounding mile.

NATURE OF IMPACT FACTOR RULES

In this application, factor rules varied from one impact factor to another (some being quite complex) and are described in detail in Janes *et al.* (1974). Many involved linear summation models, others involved neighborhood functions and local surface fits while others involved threshold calculations. By way of example the "factor rules" for rural operation impact were derived as follows:

> The rural operational factor developed and refined by M. J. Troughton is an appraisal of the rural-agricultural component of the study area focusing upon the significance of agriculture. From the potentially massive array of data that could be collected to represent and evaluate agricultural activity, the data bank contained five measures of an intermediate or mesoscale appropriate for inclusion over the whole study area and for the choice of macrocorridors. Other macroscale variables, especially those relating to individual farms, their assessed land units and enterprise characteristics, were seen as part of the detailed assessment to be performed within macrocorridors after the corridors are selected and therefore not included in the impact factor.

The five data items applicable to this factor included two elements of the Canada Land Inventory (C.L.I.), the major capability classes of the classification of soil capability for agriculture (an inherent characteristic, representing technologically feasible production potential) and the current land uses recorded by the Canada Land Inventory as the land use mapping component. Both records, taken from the 1:50,000 manuscript maps gave a range of conditions (*e.g.*, eight capability classes and five basic land uses of variable agricultural intensity).

These records were assigned ratings scaled from 8 to 1 and 5 to 1, respectively, with the highest number indicating highest value of capability and use intensity. For capability class 8 (organic soils "0") the rate was set to 4. For interpretation purposes, actual value of land use was rated on a percentage occurrence basis.

Intensity of agricultural land use was approximated by using a three-point scale derived from the value of agricultural production on a township basis. The data used were obtained from Statistics Canada for census year 1971. Study accuracy would have been enhanced if such data had been available at a finer level. While Statistics Canada can produce summary counts at the enumeration area level, the researchers were not given access to the data due to Statistics Canada rules on confidentiality. Where land use data revealed tobacco operations, the economic variable was increased to take into account the high level of capitalization in that kind of operation.

Future land use was mapped based on an interpretation of existing official plans, zoning bylaws and regional plans. Where the future land use was specified as agricultural use on a permanent or restricted basis, future land use was assessed a value of 2 or 1, respectively. This gives weight to areas recognized by planning boards as areas useful for agriculture. The values of the scores derived for each of land capability, economic value, and future land use were combined to give a combined assessment between one and five on an integer scale. This final score was the value of the agricultural operations factor.

IMPACT FACTOR ANALYSIS AND RESULTING MAPS

Impact factor calculations were performed for each individual 500-meter unit area in the study area. Map 7 shows the extent of the study area. In particular, the full data base record for a unit area was analyzed by a computer program and eight different impact values were calculated for the eight factors in the study. Each unit area was processed independently and the results passed to another program which produced a separate map for each impact factor. The maps were drawn with ink pen on an incremental digital plotter. On these maps, each 500-meter unit area was shaded individually to maintain full accuracy: no smoothing or generalization was applied.

The mapped results of impact factor analysis performed in the application are shown in Maps 8 through 15. In each map the intensity of the shading as shown on its legend indicates severity of environmental impact. Where an area is not shaded, this indicates no impact. Where the area is shaded solid black, this indicates prohibitive impact. A narrative commentary could be written on each map. To complete the example of the rural operations factor, the following is a commentary on that factor map (*i.e.*, Map 8).

The rural operational impact map shows a general dark level of shading. The study area is generally highly rated for agriculture and the map reflects this position. With some notable exceptions, the overall surface of potential impact is very high (over 80% of the study area has a heavy or severe rating). With the exception of the Indian Reserves which show the effects of low productivity and intensity of land use, all the northern parts (Middlesex, Oxford and Brant counties) except for a few river flood plain and wetland areas rate high impact. The southern areas are varied, moderate impact areas. They reflect overall lower soil capabilities. Severe impact areas reflect the combined effects of land use intensity related to income and tobacco production. This is especially the case in South

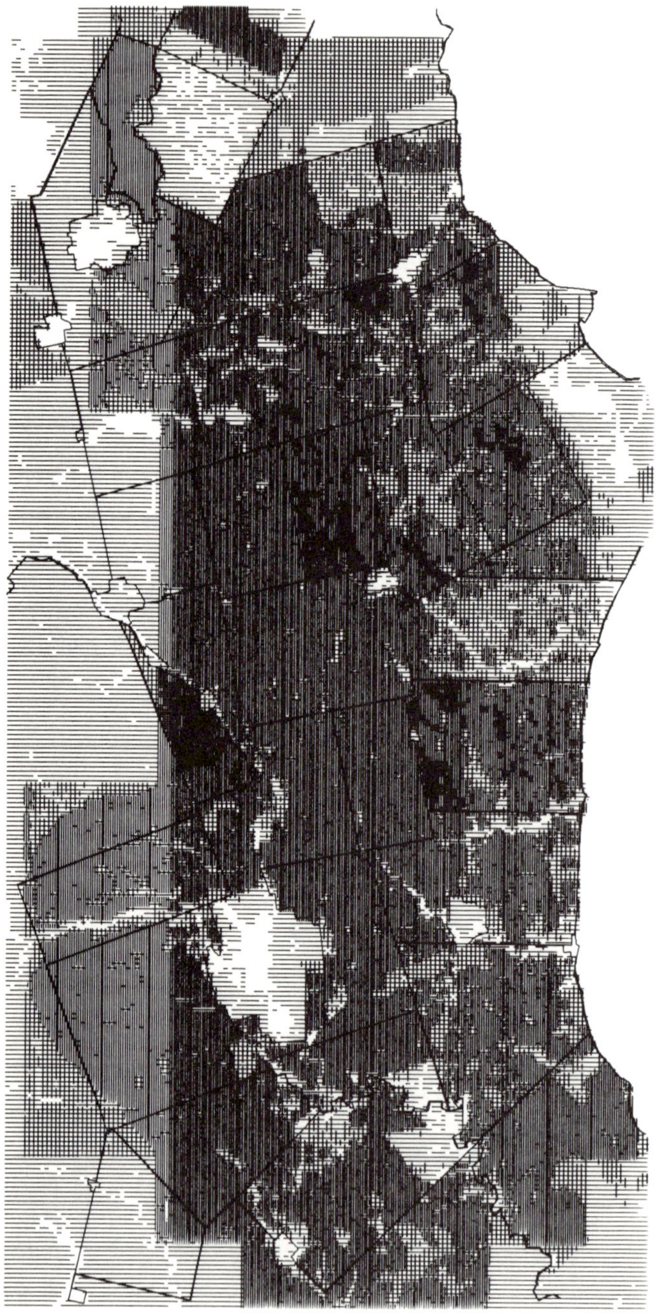

Map 8. Agriculture.

Map 9. Natural environment.

Map 10. Recreational, cultural and historical.

Map 11. Residential, institutional and commercial.

Map 12. Linear utilities.

Map 13. Area utilities and industrial.

Map 14. Natural landscape diversity.

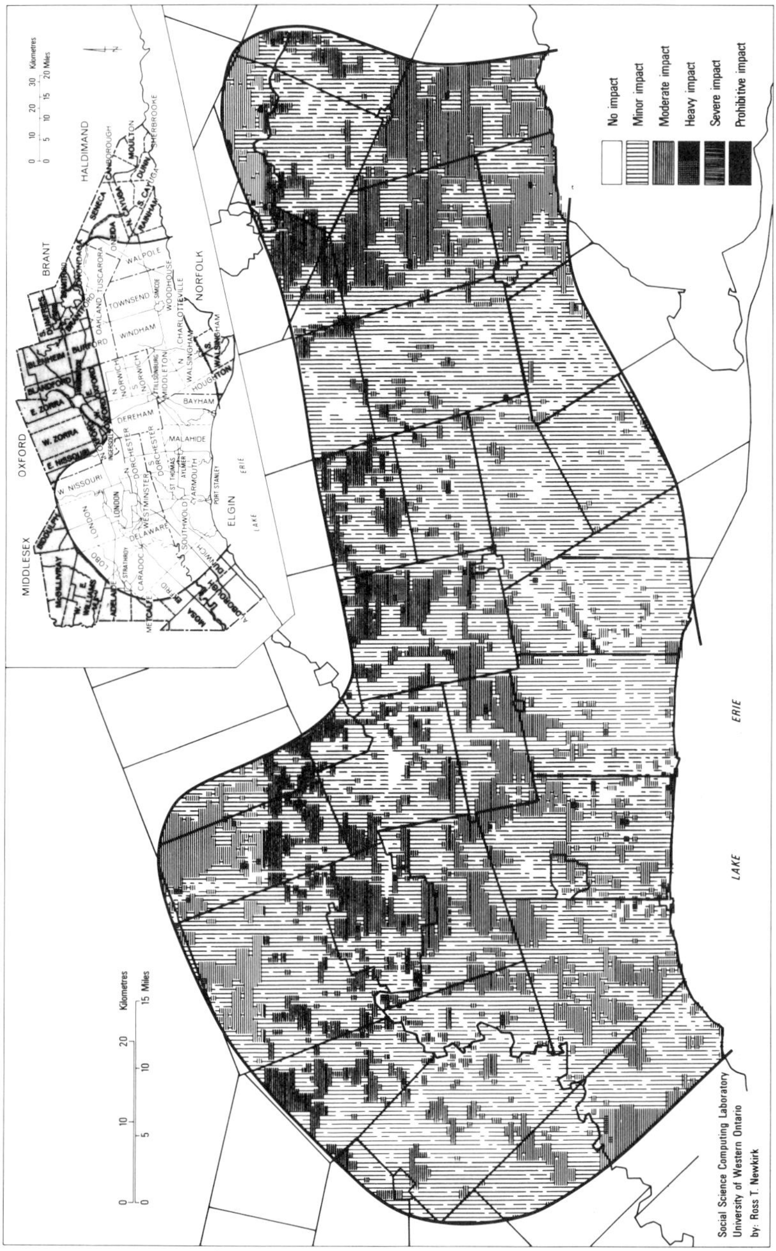

Map 15. Relative visibility.

Norwich and parts of Malahide, North Walsingham, Middleton, Windham, Charlotteville and Burford Townships. The map confirms the general character of agriculture in the area and supports the area public concern for the minimization of impact upon it.

At the same time the impact maps were produced, the public participation and local government survey results were analyzed and assessed against each factor map. The results of this comparative analysis confirmed, rather than altered the factors. No new concerns were identified with the only distinction lying in a slightly higher value placed on certain concerns, notably on agriculture and residential impacts.

COMPOSITE IMPACT AND THE CASCADE ALGORITHM

Following the calculation and mapping of the eight impact factors, the assessments were combined for each unit area to obtain an evaluation of composite impact. During the development of impact factors, major sectors of concern were isolated of quite dissimilar nature. In addition, they were established with the point of view that each factor was as important as any other in determining composite impact.

The first attempt was to use an equal weighted average to determine composite impact. The result is shown as Map 16. Since the impact factors tended to be orthogonal in nature on a unit area basis, this had the tendency of forcing composite impact to the value 3 or lower. The only time prohibitive impact could result from a combination of impact factors was if all such factors were classed prohibitive impact. A quick review of this map will show that very poor resolution is provided. The equal weight average was rejected.

The second attempt was based on the threshold concept of assigning as composite impact the highest impact observed in each unit area. This is shown in Map 17. The resolution provided by this technique is clearly far superior to that provided by the equal weight average. However, it was observed in field checks that some high impact areas were not being identified. These areas had the property that they had no major but a set of moderate impacts associated with a number of impact factors. This gave rise to the use of the cascade algorithm to provide composite impact assessment.

The third and final calculation of composite impact is shown in Map 18. The map was obtained by applying the Cascade Algorithm with base and level set equal to two. The increased discrimination provided over the threshold calculation is readily apparent. This assessment of composite impact met with wide support from the study team, Ontario Hydro

Map 16. Composite impact by weighted average.

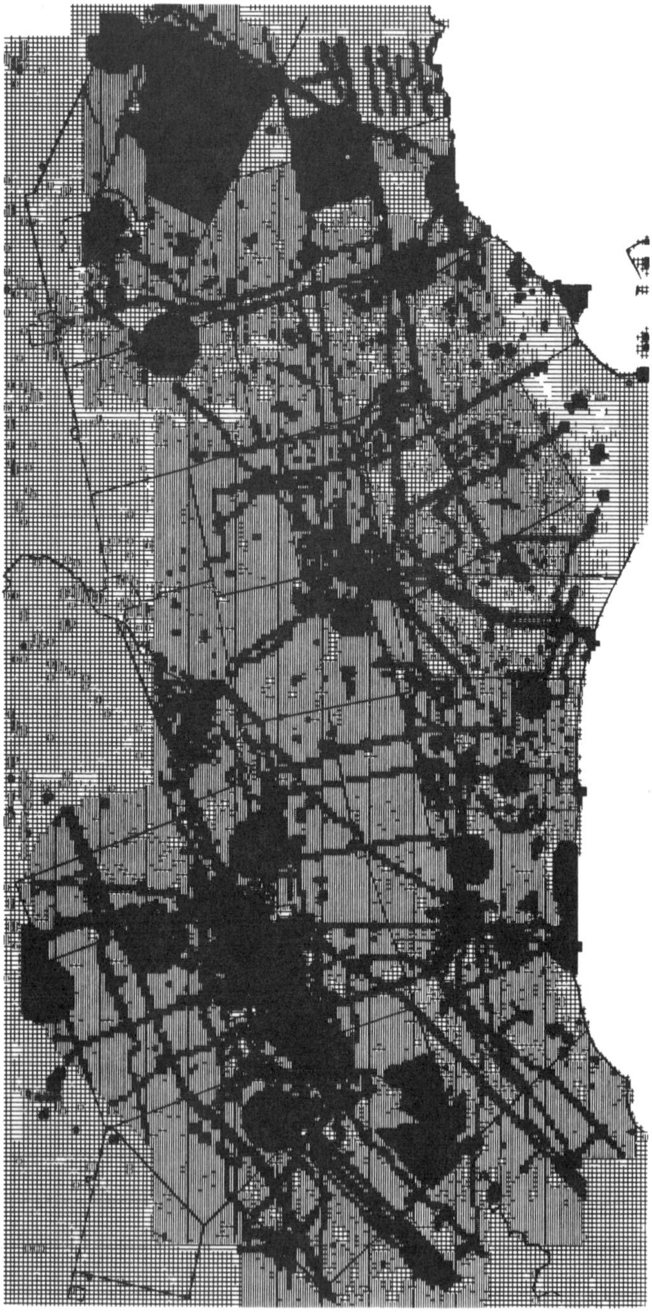

Map. 17. Composite impact by threshold.

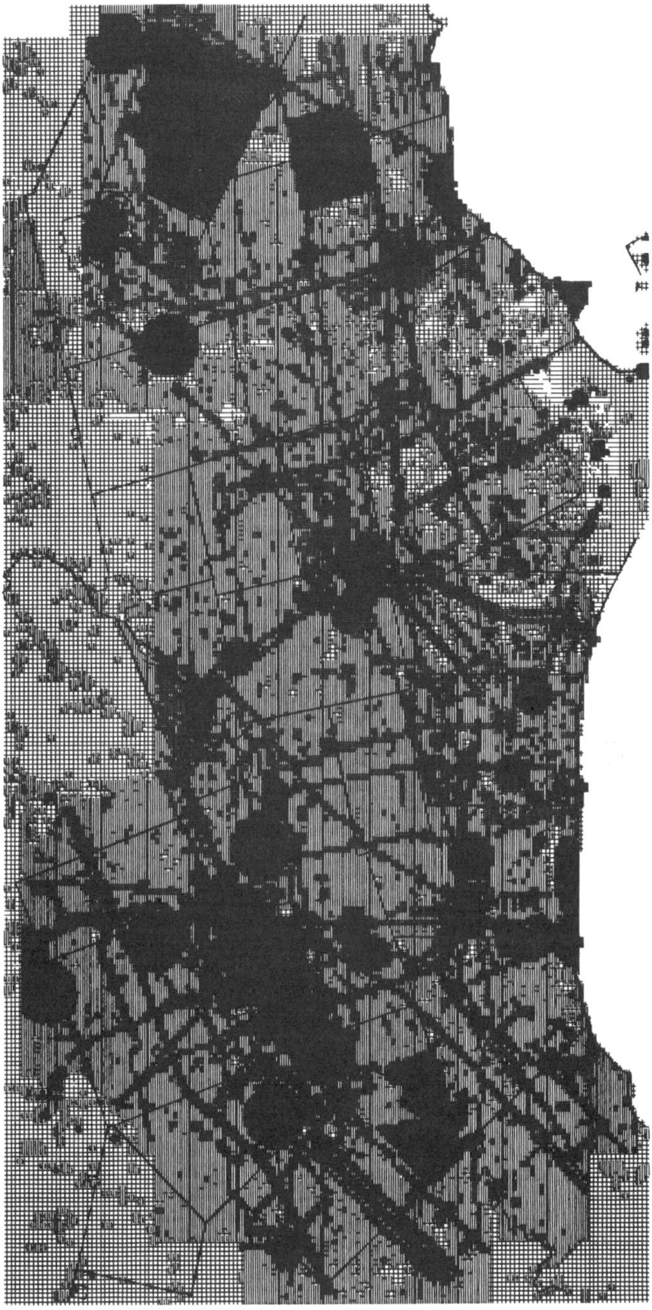

Map 18. Composite impact by Cascade Algorithm.

and government ministries. The increased resolution will permit more discriminating corridor routing analysis when the project is reactivated.

This final composite map describes a complex surface of potential impacts, some of which are mutually exclusive, and some of which overlap. The areas of greatest impact (darkest or class 6) include all residential areas, Indian reserves, special ecological sites and certain area utilities. In addition, accumulated severe impacts create a maximum impact area around the City of London and there are also more scattered areas in the prime agricultural belt from St. Thomas east through the center of the study area. Heavy and severe impact (class 4 and 5) lands are less extensive, generally reinforcing the highest impact areas. Moderate impact (class 3) areas bring in many linear utility zones, and stress particularly the broad extent of highest rated agricultural land and some top recreation areas.

Of greatest interest for corridor delimitation are the areas of slight and moderate impact (classes 2 and 3). Class 2 and 3 lands describe the largest single part of the surface. Again, they reflect particularly the high overall rural-operational impact surface. Class 2 impact lands, tend to be concentrated towards the south of the study area, but their extent is broken up by both high impact agricultural and ecological elements. Class 1 impact lands emerge as series of irregular areas, often in the more isolated rural locations. Class 1 and 2 impact lands together allow considerable scope for corridor selection away from urban areas.

SUMMARY

Three major problems were encountered during the project. This first and most difficult problem was to determine, with the level of precision required for factor calculation, the real concerns of the various ministries of the provincial government. In many cases ministry concerns and requirements are unwritten or at least unpublished. The consultants were criticized on some occasions for failing to take into account "nondocumented but implied Ministry requirements". The second problem was to assign correct values to data items and decide upon their combination and relative weights in impact factor rules. There is a paucity of reference material related to the value rating problem. The third major difficulty encountered was access to up-to-date detailed map series which provide the data required to evaluate the factors. The standard land use, agricultural, wildlife, recreation capability series dated from the mid-sixties and required substantial updates. Many valuable map series are confidential. Often these maps are very recent and detailed and could provide valuable research aides for a wide variety of problems.

Staff of James F. MacLaren Ltd., Ontario Hydro and various government ministries have reviewed the results of the recent composite impact assessment in the field. They have reported that in general a high degree of accuracy has been maintained by the analysis system across the study area. In particular, the relative visibility and natural diversity factors were subjected to a direct and careful comparison with the work performed by an independent firm of landscape architects in a very difficult sample area. The computer analysis compared very favorably with the architect's study and in places showed a higher level of sensitivity (particularly in transition areas). Analysis accuracy has been proven in the field. The system appears to be a useful tool for impact studies.

CHAPTER 12

SUMMARY OF RESULTS AND CONCLUSIONS

USEFUL FEATURES OF THE PLANNING SYSTEM

General Nature

The development of the planning system was structured independently from any specific utility or routing application. Accordingly, the system can be used for most "linear" utilities providing the appropriate data and impact factor rules are developed. Impact factor rules may be developed arbitrarily as required. The size and complexity of the data base may be developed without undue restrictions other than its implications on computer expenditures associated with its scale. The functioning and interrelations of planning system segments do not require a specific analysis scale to be used. For larger-scale studies unit areas could be 1000 or 2000 meters in size with working map scale of 1:125,000 or 1:50,000. For detailed local studies, unit areas could be 200 meters or even 50 meters in size. An important implication is that the same kind of study can be applied at various scales; this will help integrate policy and operational level planning. Application of a similar system across many studies will facilitate interproject comparisons and review.

In addition, the system can be used for other environmental location problems where routing is not important but location of a site with specific nature is. For example, location of landfill sites could be facilitated by using phases one, two and three of the planning system. Terrain data and local surface fits could be used to locate less visible and topographically suitable sites. Landform, soil texture and drainage information along with stream channel, quarry site indicators, standing water and so forth could be used to find areas least likely to spread contaminated groundwater. Other variables could be used to assess site proximity to urban centers and transportation. Approximate ecological impact factors would be developed. All of these parameters could be synthesized into a single map which would show high and low suitability for landfill sites. The map could then be searched for "suitable level" assessments which have the required spatial extent.

Computer-Assisted Data Base Development

The map sheet digitizing and subsequent interpolation of unit area characteristics has proved itself to be very accurate and productive in a significant application. Of particular value is the procedure's ability to adapt to the scale of the source document even when the document's scale differs from others in a study. This eliminates costly, error-introducing and time-consuming redrafting, scale changing, or manual interpretation.

The objective and tireless interpolation program has successfully replaced the subjective and boring manual interpretation problems of collecting unit area data from maps. During the application reported in Chapter 11, some comparisons between manual and computer-assisted data bank development were made. Results showed that data acquisition time was reduced to at least one tenth of the manual, and accuracy was increased significantly when the computer-assisted method was used.

The system utilizes relatively inexpensive digitizing equipment and requires only semiskilled operator involvement. This has resulted in a major reduction in the costs of capturing a unit of information.

An important implication is the resulting incentive to use sufficient data to perform an analysis. In addition, this new use of data input makes it reasonably possible to consider developing generalized regional planning data banks which could be used for a variety of applications.

Accommodation of Public Input

As outlined in the preceding two sections, the system itself does not place restrictions on data series to be included in a study. It could be claimed that, within reason, any data which can be mapped can be input to a study data base. It is quite possible to develop a map series based upon the results of surveys of interested groups and local governments. On such a map series, locally valued historic or cultural sites can be maintained; local naturalist study areas can be identified and so forth. Such a map series can be added to a study data base in the same way as other "regular" map series.

In addition to actual information placed in the study data base, public input can be used in impact factor algorithms. If required, a "public concerns" factor could be developed based both upon specially added "public data" and evaluations of selected study area phenomena inventoried in other data variables. Of course, public concerns could be included in the assessment of other impact factors.

Study output in map and tabular form should facilitate review by the public. Public concerns identified by the study team can be used to alter the assessment without undue difficulty. This is made possible by the "open ended nature" of the data base and the flexibility permitted in impact factor algorithms.

The current state of emerging public and governmental concern and requirements implies that criteria will change, often rapidly, during the course of a study. The planning system is structured to permit the inclusion of these and provide the flexibility to update so that studies are not obsolete before they are completed.

System-Produced Map and Tabular Output

The planning system develops high-quality pen and ink maps utilizing a computer controlled digital plotter. Certification and development of the data base, impact factor algorithm development, presentation of impact assessment, and route development are facilitated.

Redrawing the basic data input polygons for light table certification of data zones has been very effective in providing a very accurate data base. Semiskilled technicians can perform error detection and correction very quickly.

The provision of a mapping program which shades with precision each individual unit area at an arbitrarily selected scale permits display of data base variables, impact factor algorithm results and composite impact assessments. When data sources or factor or composite algorithms are altered, finished maps of the result can be obtained overnight. This permits the necessary flexibility to respond to external review.

A significant advantage of the computer-generated mapping is the facility to develop quickly and at low cost supporting documentation for a study. It is relatively easy to develop a map which is a special composition of data base variables. This could be used, for example, to display all areas involving headwater forests. Such a map might be used to support the presentation of a "natural environment" impact factor.

The most important advantage, however, is to present a series of impact factor maps and a composite map which can be readily comprehended by specialists and many nonspecialists. The computer's ability to overlay the cultural mosaic of urban boundaries, major roads and rural municipalities in conjunction with a designation of the proposed alternate corridors permits reviewers to assess quickly the integration of the utility corridor in the study area.

The tabular output which is associated with each alternate corridor permits a direct and meaningful comparison between alternatives. This not only answers major criticisms of most existing impact assessment methods, but it also assists the study team in developing the study by permitting evaluation of alternative factor, composite and routing algorithms.

Composite Impact by Cascade Algorithm

The Cascade Algorithm has been successful at permitting the combination of ordinal value impact rankings. In particular, it addresses the problem of the combination of a number of moderate value impacts which give rise to a threshold condition of higher total impact.

Increased resolution in study area composite impact analysis has been realized. This increases the sensitivity of the planning system to local discontinuities. The detection and avoidance of local higher impact values should help to provide a study which gains public and government approval and minimizes last-minute route adjustments based on field studies.

The facility to perform a good composite assessment is crucial to permitting impacts to be treated by the factor approach. The separation of impact evaluation into functional or "concern" areas facilitates study development and review since the fundamental structural components may be developed and examined separately. If this was not possible, the only alternative would be to develop one single impact index where it would be almost impossible to know what had been the major contributor to the impact rating in a specific area. The Cascade Algorithm has been indispensable to the development of the planning system.

ROUTE DEVELOPMENT

The route development process based upon a composite impact "surface" shows great promise for locating alternate minimal impact corridors for utility development, and for approaching the "no development" question. The flexibility provided during the graph contraction and adjacency calculations permits the analysis to be relaxed if required. This will permit the use of the imprecise real data that have to be used in most impact studies. In addition, it permits the study team to respond to governmental questions of the kind "if we were to relax our standards in this case, what is the next best set of routes?"

Since impact smoothing and averaging have been avoided and the basic routing algorithm is known to find minimum cost paths, the routing phase will establish the least impact corridors subject to the composite impact.

The graph contraction technique has increased the size of problems which can be analyzed. This result is general to route-finding problems and accordingly will find application in other areas such as automated computer circuit board wiring design.

SOME OUTSTANDING PROBLEMS

Three significant problems are to be faced in environmental routing development. The problems tend to be general in nature and are encountered in most environmental impact assessments.

For many significant projects, comprehensive guidelines are not available to assist in the development of impact factors. In many areas there is lack of agreement on suitable standards. The legislation is often disjointed and incomplete. Various government agencies are addressing this problem, but it is likely to be some time before this situation is rectified.

Availability of up-to-date, consistent and accurate data is a major problem. In many cases, published map series are too imprecise for planning purposes and reliance must be made on manuscript map work sheets (if they can be obtained).

Finally, a remaining difficulty is establishing ways of combining data to represent correctly the impact associated with a project. The data variables to be used, sources, and the weightings or combination techniques to be used form the "value-weighting" problem. Many study team judgments are required to implement an application of the planning system. The actual data variables to consider and the combinational rules must await much research and the development of standards. Until such time it remains imperative that environmental impact assessment methods make as explicit as possible the judgments taken in a study. In addition studies must be conducted in such a way as to facilitate adjusting the judgments.

CONCLUSION

The planning system provides a framework and analysis method which will support an environmental routing study from basic data collection, through impact analysis, to the development and assessment of alternate corridors for utility development. Its successful application on a major project established its viability.

The system is particularly well adapted to accommodate adjustments required in this current period of emerging guidelines and policy. Basic system flexibility should also serve to permit further development and applications of the planning system over time. While the system does not solve the general problems facing environment impact analysis mentioned above, it can help address them. The system forces judgments to be made explicit and provides readily comprehended (mapped and tabulated) results of such judgments. This should help establish a clearer understanding of the trade-offs involved. The facility to include public and government input both in the analysis and review stages will help clarify issues and the corresponding judgments. Finally, the structure of the planning system encourages iterative developments of an environmental routing analysis. This important feature of the system should permit studies to develop until a clear decision is possible. Indeed the process of iterating through a number of applications may help resolve some of the value-weighting problems and guideline specifications facing impact studies today.

APPENDIX I

SAMPLE STUDY DATA BASE

This appendix contains a list of the variables used in the electric transmission line study discussed in Chapter 11. It is included here not only to support the discussion in Chapter 11, but also to provide an example of the kinds of data which could be useful in environmental routing studies.

DATA BASE VARIABLES

Variable		
No.	Description	Data Codes
1	Easting military grid	Actual grid value of center of square
2	Northing military grid	Actual grid value of center of square
3	Average terrain slope	One of 0-0 to 4.9 degrees 1-5 to 14.9 degrees 2-15 and more degrees
4	Roughness	The number of 25-foot contour lines passing through the grid square boundaries
5	Standing water	0 - absent 1 - present
6	Stream channel classification	0 - no stream present 1 - first order present 2 - second order 3 - third order 4 - fourth order 5 - fifth order 6 - sixth order
7	Landform type	1 - till (end) moraine 2 - glacial spillway 3 - kame moraine 4 - till plain 5 - drumlinized till plain

Variable		
No.	Description	Data Codes
7	Landform type	6 - bevelled till plain 7 - limestone plain 8 - shale plain 9 - sand plain 10 - sand plain with dunes 11 - clay plain 12 - esker 13 - bog or marsh
8	Soil texture	10 - clay 12 - clay loam 20 - loam 24 - loamy sand 30 - silt 32 - silty loam 40 - sand 42 - sandy loam 50 - gravel 52 - gravelly loam 60 - much, peat or organic 70 - alluvial or bottom land
9	Soil drainage	1 - excessive drainage 2 - good drainage 3 - imperfect drainage 4 - poor drainage 5 - very poor drainage
10	Soil reaction	0 - unknown 1 - strong acid 2 - medium acid 3 - slightly acid 4 - neutral 5 - alkaline
11	Sand, gravel, or rock quarries	0 - absent 1 - present
12	Thickness of overburden	Number of feet from soil surface to bedrock
13	Soil capability for agriculture (main class as per ARDA coding)	1 - Class 1 2 - Class 2 3 - Class 3 4 - Class 4 5 - Class 5 6 - Class 6 7 - Class 7 8 - Class 8
14	Soil capability for agriculture (subclasses)	1 - C (as per ARDA definition) 2 - S 3 - E 4 - F 5 - I

		6 - M 7 - N 8 - P 9 - R 10 - S 11 - T 12 - W 13 - X
15	Soil capability for agriculture (secondary capability)	As described for V13
16	Soil capability for agriculture (subclass secondary)	As described for V14
17	Average elevation	In feet above sea level
18	Map identification	Index code for the 1:50,000 topographic map wherein this grid square is located
19	Township identification	township name BR - Brantford BU - Burford DU - Dunnfrees OA - Oakland O - Onandaga IR - Indian Reservation AL - Aldeborough BA - Bayham SD - Dorchester South DU - Dunwich HA - Malahide SO - Southwold YA - Yarmouth ON - Oneida WA - Wapole AD - Adelaide BI - Biddulph CA - Caradoc DE - Delaware ND - Dorchester North EK - Ekfrid LO - Lobo LON - London MC - McGillivray ME - Metcalf MO - Mosa WN - Nissouri West WE - Westminister EW - Williams East WW - Williams West CH - Charlotteville HO - Houghton MI - Middleton TO - Townsend

Variable No.	Variable Description	Data Codes
19	Township identification	NW - Walsingham North SW - Walsingham South WI - Windham WO - Woodhouse B - Blandford BLE - Blenheim DER - Dereham EN - Nissouri East NN - Norwich North SN - Norwich South EO - Oxford East NO - Oxford North WO - Oxford West EZ - Zorra East WZ - Zorra West BL - Blanchard
20	First ranking land use (smallest item detectable 6.25% of grid square or 3.75 acres)	B - built-up area E - mine, quarry, sand or gravel pit O - outdoor recreation HG - intensive agriculture A - 75% to 100% cropland P - 25% to 74.9% cropland A2P2 - 74.9% to 100% improved pasture K - rough grazing land and rangeland TU - woodland
21	Grid square coverage for V20	Percentage
22	Second ranking land use	As described for V20
23	Grid square coverage for V22	Percentage
24	Third ranking land use	As described for V20
25	Grid square coverage for V24	Percentage
26	Woodlot indicator	TREE - present NULL - absent
27	Grid square coverage for woodlot	Percentage
28	First ranking corrected land use (smallest item detectable 11.11% of grid square or 6.67 acres)	IA - intensive agriculture RC - rural residential clusters UR - urban residential CO - commercial IN - industrial TK - tobacco kilns ST - institutional EX - experimental farm
29	Grid square coverage for V28	Percentage
30	Second ranking corrected land use	As described for V28
31	Grid square coverage for V30	Percentage
32	Third ranking corrected land use	As described for V28
33	Grid square coverage for V32	Percentage

34	Tobacco kilns	TK - present blank - absent
35	Indian reservations	IR - present blank - absent
36	First ranking zoning (smallest item detectable 25% of grid square or 15 acres)	R - urban residential C - commercial INS - institutional I - industrial CON - open space, recreation, conservation, flood plain or hazard land A - agricultural and rural AR - restricted agriculture EST - estate residential PQ - pits and quarries DVT - optional development land CP - proposed conservation authority acquisitions NC - not classified
37	Grid square coverage by first ranking zoning	Percentage
38	Second ranking zoning	As described for V36
39	Grid square coverage by second ranking zoning	Percentage
40	Third ranking zoning	As described for V36
41	Grid square coverage by third ranking zoning	Percentage
42	First ranking utility (smallest item detectable 11.11% of grid square or 6.67 acres)	GP - gas pipeline OP - oil pipeline HP - proposed highway corridor HA - approved new highway SL - sewage lagoon RR - railroad, operating RA - railroad, abandoned AP - air field, licensed AF - air field, unlicensed AA - air field, abandoned HM - radio transmitter RM - radio transmitter, AM or FM TV - television transmitter NB - radar, microwave, navigation beacon CT - cable TV antenna OHR - Ontario Hydro major right-of-way (owned) OHE - Ontario Hydro major right-of-way (easement) LQ - sanitary landfill, opened LC - sanitary landfill, closed LP - sanitary landfill, proposed WP - water pipeline (existing) PWP - water pipeline (proposed)

Variable No.	Variable Description	Data Codes
42	First ranking utility (smallest item detectable 11.11% of grid square or 6.67 acres)	IN - point of interest NC - not classified
43	Grid square coverage by first ranking utility	Percentage
44	Second ranking utility	As described for V42
45	Grid square coverage of second ranking utility	Percentage
46	Third ranking utility	As described for V42
47	Grid square coverage of third ranking utility	Percentage
48	First ranking natural resource (smallest item detectable 11.11% of grid square or 6.67 acres)	PR - conservation lands FI - fishery CF - county forest CL - crown land W - W.I.A. forest and management area Q - gravel pits and quarries M - McIlwraith Field Naturalist study area PC - campgrounds IV - Indian Village IP - Indian palisade RH - regionally historic LH - locally historic NC - not classified
49	Grid square coverage by first ranking natural resource	Percentage
50	Second ranking natural resource	As described for V48
51	Grid square coverage by second ranking natural resource	Percentage
52	Third ranking natural resource	As described for V48
53	Grid square coverage by third ranking natural resource	Percentage
54	First ranking ungulate capability (smallest item detectable 25% of grid square or 15 acres)	Full A.R.D.A. region code giving seven (7) levels of capability and limitation support codes as per maps
55	Grid square coverage by first ranking ungulate capability	Percentage
56	Second ranking ungulate capability	As described for V54
57	Grid square coverage by second ranking ungulate capability	Percentage
58	Third ranking ungulate capability	As described for V54
59	Grid square coverage by third ranking ungulate capability	Percentage
60	First ranking waterfowl capability (smallest item detectable 25% of grid square or 15 acres)	Full A.R.D.A. region code giving seven (7) levels of capability and limitation support codes as per maps

61	Grid square coverage by first ranking waterfowl capability	Percentage
62	Second ranking waterfowl capability	As described for V60
63	Grid square coverage by second ranking waterfowl capability	Percentage
64	Third ranking waterfowl capability	As described for V60
65	Grid square coverage by third ranking waterfowl capability	Percentage
66	First ranking capability for outdoor recreation (smallest item detectable 25% of grid square or 15 acres)	Full A.R.D.A. region code giving seven (7) levels of capability and subclasses as per maps
67	Grid square coverage by first ranking outdoor recreation	Percentage
68	Second ranking capability for outdoor recreation	As described for V66
69	Grid square coverage by second ranking outdoor recreation	Percentage
70	Third ranking capability for outdoor recreation	As described for V66
71	Grid square coverage by third ranking outdoor recreation	Percentage
72	Historic site indicator	Percentage of grid
73	Average elevation of local surface	Feet above sea level
74	Average woodlot coverage in surrounding mile	Percentage
75	Average roughness in surrounding mile	Average (see V4)
76	Atlantic Flyway and Hawk's Cliff indicator	ATL - Atlantic Flyway HAWK - Hawk's Cliff MISS - Mississippi Flyway NC - absent
77	Mississippi Flyway indicator	1 - present 0 - absent
78	Timber use capability	Seven levels of capability
79	First ranking natural environment feature (smallest item detectable 11.11% of grid square or 6.67 acres) Designated by the Ministry of Natural Resources	*Sensitive Areas* S1 - Dorchester Swamp S2 - Thames River Floodplain S3 - Melbourne Marsh S4 - Sifton (Byron) Bog S5 - Westminister Ponds S6 - Foster Ponds S7 - Dingman Ponds S8 - Norwich Junction Woods S9 - Vienna Pawpaw Stand S10 - Jolly Swamp S11 - Catfish Creek Sand Slope and Floodplain S12 - Monroe London's Woodlot S13 - Oak Parkland S14 - Backus Woods

No.	Variable Description	Data Codes
79	First ranking natural environment feature (smallest item detectable 11.11% of grid square or 6.67 acres) Designated by the Ministry of Natural Resources	S15 - Spooky Hills Woodland S16 - Bert Arthur's Mixed Hardwoods S17 - Black Spruce Stand S18 - Spooky Hollow Nature Sanctuary S19 - Marburg Station Swamp S20 - Charlotteville Rolling Sandland S21 - Walsh Rolling Sandland S22 - South Walsingham Sand Ridges S23 - Windham Center Sandy Swamp-land S24 - Teeterville Sand Ridges S25 - Springvale Swamp S26 - Nanticoke Creek Marsh *Historic Sites* HP - provincially significant HR - regionally significant HL - locally significant HH - historically significant *Stream Quality* QS - spawning area QH - high quality QHL - high quality with limitations QM - medium quality *Proposed Areas* PCA - proposed conservation area PFA - proposed forestry area PPP - proposed provincial park NC - not classified
80	Grid square coverage for first ranked natural environment features	Percentage
81	Second ranking natural environment feature	As described for V79
82	Grid square coverage for second ranked natural environment feature	Percentage
83	Third ranking natural environment feature	As described for V79
84	Grid square coverage for third ranked natural environment feature	Percentage
85	Sensitive areas	S - present - absent
86	First ranking wildlife suitability (smallest item detectable 25% of grid square or 15 acres)	Ontario Wildlife Suitability Inventory, seven (7) levels of suitability for each of deer or ruffed grouse, pheasant or Hungarian partridge, geese and water-fowl

87	Grid square coverage first ranking wildlife suitability	Percentage
88	Second ranking wildlife suitability	As described for V86
89	Grid square coverage for second ranking wildlife suitability	Percentage
90	Third ranking wildlife suitability	As described for V86
91	Grid square coverage for third ranking wildlife suitability	Percentage

REFERENCES

Akers, S. B. "A Modification of Lee's Path Connection Algorithm," *IEEE Trans. Electron. Comp.*, EC-16:97,98 (February 1967).

Aldred, B. K. *Point in Polygon Algorithms.* Report No. 25, UK. (Peterlee, County Durham: Scientific Center, IBM United Kingdom Ltd., undated.)

Alexander, C., and M. L. Manheim. "The Use of Diagrams in Highway Route Location: An Experiment," *Civil Engineering Systems Laboratory Research Report R62-3* (Cambridge, MA: Massachusetts Institute of Technology, 1962).

Andersen, S. O. "Concepts and Methods of Benefit-Cost Analysis," in *Environmental Impact Assessment: Guidelines and Commentary,* T. G. Dickert and K. R. Domeny, Eds. (Berkeley, CA: University of California, 1974), pp. 89-105.

Anderson, G. D., S. P. Y. Lee and J. W. Bush. "CCISS: Conversational Computer Information and Statistical System, User's Reference Manual," mimeograph, Health Sciences Computing Service, McMaster University, Hamilton, Ontario (undated).

Aramaki, I., T. Kawabata and K. Arimoto. "Automation of Etching-Pattern Layout," *Comm. ACM* 14(11):720-730 (1971).

Barto, W. P. "The Agricultural, Forestry, Recreational and Wildlife Opportunity Cost of Pipelines, Hydro Lines and Highways," Masters Thesis, Natural Resources Institute, University of Manitoba, Winnipeg (1974).

Berztiss, A. T. *Data Structures Theory and Practice* (New York: Academic Press, Inc., 1971).

Bishop, R. C. "Uncertainty and the Evaluation of Environmental Impacts: A Cast Study," in *Environmental Impact Assessment: Guidelines and Commentary,* T. G. Dickert and K. R. Domeny, Eds. (Berkeley, CA: University of California, 1974), pp. 107-120.

Burnham, J. B., M. H. Karr, G. L. Wilfert, W. S. Maynard and S. M. Nealey. *A Technique for Environmental Decision Making Using Quantified Social and Aesthetic Values* (Richland, WA: Battelle Pacific Northwest Laboratories, 1974).

Caldwell, T. "On Finding Minimum Routes in a Network with Turn Penalties," *Comm. ACM* 4:107,108 (1961).

Principles for Environmental Impact Assessment (Toronto: Canadian Environmental Law Association, 1973).

Cederbaum, I. "Optimal Backboard Ordering through the Shortest Path Algorithm," *Proc. IEEE International Symposium on Circuits and Systems* (1974), pp. 186-190.

Cooper, L., and D. Steinberg. *Introduction to Methods of Optimization* (Toronto: W. B. Saunders Company, 1970).

Denenberg, S. A., C. C. Corbin and P. A. Alsberg. *NARIS, National Resource Information System.* (Champaign-Urbana, IL: Center for Advanced Computation, University of Illinois, undated).

Dickert, T. G. "Methods for Environmental Impact Assessment: A Comparison," in *Environmental Impact Assessment: Guidelines and Commentary* (Berkeley, CA: University of California, 1974), pp. 127-143.

Dickert, T. G., and J. C. Sorensen. "Some Suggestions on the Content and Organization of Environmental Statements," in *Environmental Impact Assessment: Guidelines and Commentary* (Berkeley, CA: University of California, 1974), pp. 35-53.

Dickert, T. G., and K. R. Domeny, Eds. *Environmental Impact Assessment: Guidelines and Commentary* (Berkeley, CA: University of California, 1974).

Dijkstra, E. W. "A Note on Two Problems in Connection with Graphs," *Numer. Math.* 1:269-271 (1959).

Ditton, R. B., and T. I. Goodale, Eds. *Environmental Impact Analysis: Philosophy and Methods* (Green Bay, WI: University of Wisconsin, 1972).

An Environmental Assessment of Nanaimo Port Alternatives. (Ottawa: Environment Canada, Outdoor Recreation–Open Space Division, Lands Directorate, 1974).

Fisher, D. W., and G. S. Davies. "An Attempt to Make Environmental Evaluation Systematic and Explicit," in *National Conference on Environmental Impact Assessment: Philosophy and Methodology*, C. G. Morley and B. Odlum, Eds. (Winnipeg: University of Manitoba, Agassiz Center for Water Studies, 1973), pp. 155-187.

Fisk, C. J., D. L. Caskey and L. E. West. "Topographic Simulation as an Aid to Printed Circuit Board Design," in *Proceedings of the SHARE Design Automation Workshop* 4:17.1-17.23 (1967).

Freeman, H. "Computer Processing of Line-Drawing Images," *ACM Comp. Surveys* 6(1):57-97 (1974).

Fukunaga, K., and P. M. Narendra. "A Branch and Bound Algorithm for Computing k-Nearest Neighbours," *IEEE Trans. Comp.* CC-24(7):750-753 (1975).

Goodchild, M. F. "Error Estimation for Geographic Data Structures," mimeograph, Department of Geography, University of Western Ontario, London (undated).

Goodchild, M. F. "A Position Paper on Geographic Data Processing," Report to the Department of Indian and Northern Affairs (contract 72-114), National and Historic Parks Branch, Department of Geography, University of Western Ontario, London (1973).

Goodchild, M. F. "A Program Package for Handling Selected Geographic Data Structures," mimeograph, Department of Geography, University of Western Ontario, London (1973).

Gordon, P., and W. K. MacReynolds. "A Quadratic Assignment Approach to Urban Land Use: Optimization Algorithms for Different Size Problems with Unknown Statistical Properties of the Criterion Function," *Proceedings of the 8th Annual ACM Urban Symposium* (New York: Association for Computing Machinery, 1973).

Hart, P. E., N. J. Nilsson and B. Raphael. "A Formal Basis for the Heuristic Determination of Minimum Cost Paths," *IEEE Trans. Systems Sci. Cybernet* SSC-4(2):100-107 (July 1968).

Hausmanis, I. "Computer Aided Analysis of Alternative Route Locations within the Guelph-Dundas Test Area," M. Eng. Thesis, Department of Civil Engineering, University of Toronto (1971).

Hill, M., and Y. Tzamir. "Multidimensional Evaluation of Regional Plans Serving Multiple Objectives," *Papers Reg. Sci. Assoc.* 29:139-165 (1972).

Hirvonen, H. E., J. P. Senyk and E. T. Oswald. *Saltspring Island, A Landscape Analysis,* Report BC-X-99, Pacific Forest Research Center, Canadian Forestry Service (Ottawa: Department of the Environment, 1974).

Hoffman, D. W. *The Assessment of Soil Productivity for Agriculture, ARDA Report Number 4,* Canada Land Inventory Program (Ottawa: Department of the Environment, 1971).

Isard, W., and R. V. Van Zele. "Practical Regional Science Analysis for Environmental Management," *Int. Reg. Sci. Rev.* 1(1):1-25 (1975).

Jackson, J. N., and J. Forrester. *Practical Geography* (Toronto: McGraw-Hill Ryerson, 1974).

Janes, S. H., A. R. Glasgow, J. E. Dooley, R. T. Newkirk, M. J. Troughton, I. Moncrieff and D. Gagel. *Interim Report: Environment Study for an Electric Power Transmission Right-of-Way,* 2 Vol. (Toronto: James F. MacLaren Limited, 1974).

Jankulak, F. J. *A Systems Design for a Geographic Information System,* Publication Z77-7121. Technical Information Exchange, IBM internal use only (White Plains, NY: International Business Machines, 1967).

Johnson, A. I. "Computer Aided Process Analysis and Design–A Modular Approach," *Brit. Chem. Eng. Proc. Tech.* 17(1):28-33 (1972).

Jowett, D. "The Quantitative Assessment of Environmental Impacts," in *Environmental Impact Analysis: Philosophy and Methods,* R. B. Ditton and T. I. Goodale, Eds. (Green Bay, WI: University of Wisconsin, 1972), pp. 127-136.

Kimme, C., D. Ballard and J. Sklansky. "Finding Circles by an Array of Accumulators," *Comm. ACM* 18(2):120-122 (1975).

Kirby, R. F. "The Minimum Route Problem for Networks with Turn Penalties and Prohibitions," *Trans. Res.* 3:397-408 (1969).

Klaassen, L. H., and T. H. Botterweg. "Evaluating a Socioeconomic and Environmental Project," *Papers Reg. Sci. Assoc.* 33:155-175 (1974).

Knuth, D. E. *The Art of Computer Programming: Fundamental Algorithms,* Vol. 1, 2nd ed. (Don Mills, Ontario: Addison-Wesley, Ltd, 1969).

Knuth, D. E. *The Art of Computer Programming: Sorting and Searching,* Vol. 3 (Don Mills, Ontario: Addison-Wesley, Ltd., 1973).

Krauskopf, T. M., and D. C. Bunde. "Evaluation of Environmental Impact Through a Computer Modelling Process," in *Environmental Impact Analysis: Philosophy and Methods,* R. B. Ditton and T. I. Goodale, Eds. (Green Bay, WI: University of Wisconsin, 1972).

Kreith, F. "Lack of Impact," *Environment* 15(1):26-33 (1973).

Lee, C. Y. "An Algorithm for Path Connections and its Applications," *IRE Trans. Electron. Comp.* EC-10:346-365 (September 1961).

Leopold, L. B., F. E. Clarke, B. B. Hanshaw and J. R. Balsley. *A Procedure for Evaluating Environmental Impact,* Geological Survey Circular 645 (Washington, DC: U.S. Department of Interior, 1971).

Light, R. A., and R. C. Johnson. "Similarity Between Computer Control Systems and Community Planning," *Simulation* (July 1974), pp. 1-9.

Lindgren, D. T. "Applications of Remote Sensing to the Monitoring of Urban Growth," *Pub. Data Use* 3(3):2-6 (1975).

Lucas, A. R. "Environmental Impact Assessment: Legal Perspective," in *National Conference on Environmental Impact Assessment: Philosophy and Methodology,* C. G. Morley and B. Odlum, Eds. (Winnipeg: University of Manitoba, Agassiz Center for Water Studies, 1973), pp. 30-60.

Lyle, J., and M. Von Wodtke. "An Information System for Environmental Planning," *AIP J.* (November 1974), pp. 394-413.

Malisz, Boleslaw. "Threshold Analysis as a Tool in Urban and Regional Planning," *Papers Reg. Sci. Assoc.* 29:168-177 (1972).

Mastenbroek, A. P., and P. Nijkamp. "Environmental Protection and Spatial Allocation of Investments," *Int. Reg. Sci. Rev.* 1(1):73-86 (1975).

McDaniel, R., A. K. Philbrick, V. W. Sim, H. W. Taylor, M. J. Troughton and W. H. Wightman. *The Erie Research Project: Construction of a Multi-Purpose Data Bank* (London, Ontario: University of Western Ontario, 1970).

McHarg, I. L. *Design with Nature.* American Museum of Natural History (New York: Doubleday/Natural History Press, 1971).

McLellan, A. G. "Aerial Photographs–Their Manipulation and Application," in *Practical Geography,* J. N. Jackson and J. Forrester, Eds. (Toronto: McGraw-Hill Ryerson, 1974).

McNeil, J. "Environmental Impact Assessment: Economic Perspective," in *National Conference on Environmental Impact Assessment: Philosophy and Methodology,* C. G. Morley and B. Odlum, Eds. (Winnipeg: University of Manitoba, Agassiz Center for Water Studies, 1973), pp. 61-71.

McTaggart-Cowan, I. "A Methodology for Prediction of Environmental Effects," in *National Conference on Environmental Impact Assessment: Philosophy and Methodology* (Winnipeg: University of Manitoba, Agassiz Center for Water Studies, 1973), pp. 108-132.

Merrill, R. D. "Representation of Contours and Regions for Efficient Computer Search," *Comm. ACM* 16(2):69-82 (1973).

Miernyk, W. H. *The Elements of Input-Output Analysis* (New York: Random House, Inc., 1965).

Montanari, U. "A Note on Minimal Length Polygonal Approximation to a Digitized Contour," *Comm. ACM* 13(1):41-47 (1970).

Morley, C. G., and B. Odlum, Eds. *National Conference on Environment Impact Assessment: Philosophy and Methodology* (Winnipeg: University of Manitoba, Agassiz Center for Water Studies, 1973).

Naylor, T. H., J. L. Balintfy, D. S. Burdick and K. Chu. *Computer Simulation Techniques* (New York: John Wiley & Sons, Inc., 1968).

Neiman, B. J., and A. H. Miller. *Interstate 57: An Application of Computers to Highway Location Dynamics* (Green Bay, WI: University of Wisconsin, Environmental Awareness Center, undated).

Newkirk, R. T., and J. E. Dooley, "A Planning System to Minimize Environmental Impact Applied to Route Selection," *Kybernetes,* Vol. 5:213-230 (1976).

Newkirk,R. T., and M. J. Troughton. "Computer Based Environmental Assessment Applied to Electricity Transmission Corridors," *Geographical Inter-University Resource Management Seminars,* Vol. 5:31-45 (Waterloo, Ontario: Wilfrid Laurier University, Department of Geography, 1975).

Nijkamp, P., and J. H. P. Paelinck. "An Interregional Model of Environmental Choice," *Papers Reg. Sci. Assoc.* 31:51-72 (1973).

Green Paper on Environmental Assessment (Toronto: Ontario Ministry of the Environment, 1973).

Orloff, N. "Suggestions for Improvement of the Environmental Impact Statement Problem," in *Environmental Impact Analysis: Philosophy and Methods,* R. B. Ditton and T. L Goodale, Eds. (Green Bay, WI: University of Wisconsin, 1972), pp. 29-41.

Peucker, T. K., and N. Chrisman. "Cartographic Data Structures," mimeograph, Department of Geography, Simon Fraser University, Burnaby (undated).

Peuker, T. K. "Computer Cartography," Resource Paper No. 17, Association of American Geographers, Commission on College Geography, Washington, DC (1972).

Plewes, M. E., and D. R. Young. "Environmental Impact Assessment: Government Perspective," in *National Conference on Environmental Impact Assessment: Philosophy and Methodology,* C. G. Morley and

B. Odlum, Eds. (Winnipeg: University of Manitoba, Agassiz Center for Water Studies, 1973), pp. 1-21.

Pohl, I. "A Theory of Bi-Directional Search in Path Problems," in *IBM Research,* limited distribution, International Business Machines, Thomas J. Watson Research Center, Yorktown Heights, New York (1969).

Popek, J., and P. Kingston. "Spatial Data File Structure Considerations," in *Geographic Data Handling,* R. F. Tomlinson, Ed. (Ottawa: International Geographical Union Commission on Geographic Data Sensing and Processing, 1972), pp. 867-889.

Potts, J. "Concept and Location Algorithm for a Multi-Purpose Communications Corridor." MA Thesis, Department of Geography, University of Western Ontario, London (1975).

Rattray, A. B., R. M. Grodzik and H. E. Jackson. "Physical Impact and Planning in the Water Landscape," in *National Conference on Environmental Impact Assessment: Philosophy and Methodology* (Winnipeg: University of Manitoba, Agassiz Center for Water Studies, 1973), pp. 208-220.

Resources for the Future. "Environmental Matrix," from the 1972 Annual Report, Resources for the Future, Inc., Washington, DC (1972).

Robinette, C. O. *Energy and Environment* (Dubuque, IA: Kendall/Hunt Publishing Company, 1973).

Ross, J. H. *Quantitative Aids to Environmental Impact Assessment* (Ottawa: Environment Canada, Lands Directorate, 1974).

Scott, A. J. *Combination Programming, Spatial Analysis and Planning* (London: Methuen & Co. Ltd, 1971).

Scott, S., and J. G. Nelson. *Outdoor Environmental Education in Canada: A Preliminary Study* (Toronto: National and Provincial Parks Association of Canada, 1973).

Sewell, W. R. D., and I. Burton, Eds. *Perceptions and Attitudes in Resources Management* (Ottawa: Department of Energy, Mines and Resources, Policy Research and Coordination Branch, 1971).

Steiner, D. "Data Collection by Remote Sensing," in *Geographic Data Handling,* R. F. Tomlinson, Ed. (Ottawa: International Geographical Union Commission on Geographic Data Sensing and Processing, 1972), pp. 118-353.

Steiner, D., and T. Stanhope. "Data Needs and Data Specification," in *Geographic Data Handling* (Ottawa: International Geographical Union Commission on Geographical Data Sensing and Processing, 1972), pp. 47-61.

Stigall, P. D., and O. Tasar. "A Review of Directed Graphs as Applied to Computers," *Computer* 7(10):39-47 (1974).

Stinton, D. "Problems of Grid and Boundary Overlays," in *Geographic Data Handling,* R. F. Tomlinson, Ed. (Ottawa: International Geographical Union Commission on Geographical Data Sensing and Processing, 1972), pp. 773-775.

Stinton, D., V. LaGarde and P. Kingston. "Overlay of Points on Areas," in *Geographic Data Handling,* R. F. Tomlinson, Ed. (Ottawa: International Geographical Union Commission on Geographical Data Sensing and Processing, 1972), pp. 805-818.

Sutterlin, P. G., and M. A. Cooper. "SAFRAS: A Generalized Data Base Management System," mimeograph, Department of Geology, University of Western Ontario, London (undated).

Tobler, W. R. "Selected Computer Programs," mimeograph, Department of Geography, University of Michigan, Ann Arbor (1970).

Tomlinson, R. F., Ed. *Geographic Data Handling* (Ottawa: International Geographical Union Commission on Geographic Data Sensing and Processing, 1970).

Turner, A. K. *Computer-Assisted Regional Corridor Selection* (Toronto: University of Toronto, Department of Civil Engineering, 1970).

Turner, A. K. "GCARS Generalized Computer-Aided Route Selection," mimeograph, Department of Civil Engineering, University of Toronto (1972).

Twiss, R. H. "Linking the EIS to the Planning Process," in *Environmental Impact Assessment: Guidelines and Commentary,* T. G. Dickert and K. R. Domeny, Eds. (Berkeley, CA: University of California, 1974), pp. 5-16.

U.S. Congress. *An Act to Establish a National Policy for the Environment, to Provide for the Establishment of a Council on Environmental Quality, and for Other Purposes,* 91st Congress, S.1075, January 1, 83 stat. 852 (1970).

U.S. Department of Agriculture. "Rural Electrification Administration Soil Conservation Service Environmental Impact Statements," *Federal Register* 38(222) (November 19, 1973).

U.S. Department of the Interior, and U.S. Department of Agriculture. *Environmental Criteria for Electric Transmission Systems.* (Washington, DC: U.S. Government Printing Office, 1970).

U.S. Environmental Protection Agency. *Environmental Impact Statement Guidelines,* rev. ed. (Seattle, WA: U.S. Environmental Protection Agency, 1973).

Usher, P. J. "Environmental Impact Assessment: Social Perspective," in *National Conference on Environmental Impact Assessment: Philosophy and Methodology,* C. G. Morley and B. Odlum, Eds. (Winnipeg: University of Manitoba, Agassiz Center for Water Studies, 1973), pp. 72-90.

Vaughn, A. V. "A Background Paper Prepared for the Workshop on Visual Impact," mimeograph, Forestry Department, Ontario Hydro, Toronto (1973).

Weiner, P., and E. J. Deak. *Environmental Factors in Transportation Planning* (Toronto: Lexington Books, D. C. Heath & Company, 1972).

Williams, Robin. "A Survey of Data Structures for Computer Graphic Systems," *Comp. Surveys* 3(1):1-22 (1972).

Williams, T. A., and G. P. White. "A Note on Yen's Algorithm for Finding the Length of All Shortest Paths in n-Node Nonnegative-Distance Networks," *JACM* 20(3):389,390 (1973).

Yen, J. Y. "Finding the Lengths of All Shortest Paths in n-Node Nonnegative-Distance Complete Networks Using 1/2 N Cubed Additions and N Cubed Comparisons," *JACM* 19(3):423,424 (1972).

INDEX